THE PRESERVATION OF OUR SCENERY

The Spire of Norwich Cathedral

THE PRESERVATION OF OUR SCENERY

Essays and Addresses

by
VAUGHAN CORNISH, D.Sc.
with Illustrations by the Author

CAMBRIDGE
AT THE UNIVERSITY PRESS
1937

CAMBRIDGE
UNIVERSITY PRESS

University Printing House, Cambridge CB2 8BS, United Kingdom

Cambridge University Press is part of the University of Cambridge.

It furthers the University's mission by disseminating knowledge in the pursuit of education, learning and research at the highest international levels of excellence.

www.cambridge.org
Information on this title: www.cambridge.org/9781107492806

First published 1937
First paperback edition 2015

A catalogue record for this publication is available from the British Library

ISBN 978-1-107-49280-6 Paperback

TO

SIR GUY DAWBER, R.A.

Vice-President of the Council for the Preservation of Rural England

With esteem and regard
I dedicate this book
V. C.

CONTENTS

CONTENTS

ILLUSTRATIONS

PREFACE

The beauty of scenery is no mere luxury of pleasant living but a factor in the formation of national character and its preservation is essential to the national well-being. The influence of natural beauty upon personality is moreover increasing daily. We have long since cast aside the mediaeval doctrine that Nature was the abode of evil spirits and are now shaking off the habit of studying Nature solely for the elucidation of its mechanism. Thus at long last we are beginning to cultivate that habit of receptive contemplation of natural beauty which is a source of religious inspiration of far greater potency than is yet commonly realised.

VAUGHAN CORNISH

Inglewood, Gordon Road
Camberley, Surrey
August 31st, 1936

ACKNOWLEDGMENTS

The author returns thanks to the editors of *The Tree Lover*, *Geography*, *South-Eastern Naturalist and Antiquary*, and *The Geographical Journal* for permission to reproduce papers and articles included in this volume.

CHAPTER I

NATIONAL PARKS IN ENGLAND & WALES

1. TYPES OF SCENERY SUITABLE FOR THE NATIONAL PARKS OF ENGLAND AND WALES, WITH LOCAL EXAMPLES

(*Evidence given on behalf of the C.P.R.E. at a Meeting of the National Park Committee, appointed by the Prime Minister, Dec. 3rd, 1929*)

The purpose of national parks, stated in the terms of reference, is twofold, preservation of natural character and recreation for the people, and it appears to be intended that every park should satisfy both conditions.

The scenery of England and Wales exhibits several well-marked types owing to the variety of geological structure. Therefore no scheme of national parks can adequately preserve the natural character of those parts of the country which are still in a wild or open state unless there be at least one for each principal type of scenery. These types are coast scenery, mountain scenery, moor and down, river gorges, woodland and fenland (in the original sense), six in all. The whole area of enclosed agricultural land is unsuitable for national parks. Here the public must scatter for holidays, not concentrate in a kind of country where standing crops cramp free movement. In the selection of particular areas two principles should be kept in mind. (1) The areas should be pre-eminent in beauty and, if possible, have advantages of climate. (2) They should be distributed as equitably as possible in relation to the chief industrial districts, which may be taken as centred in London, Cardiff, Birmingham, Manchester with Leeds, and Newcastle-upon-Tyne. A district of wild scenery which is immediately adjacent to a great town is more suitable for the expenditure of municipal than national funds. The areas cited in the report as suitable are such as would be mainly visited for a sojourn during the holiday season, not on day trips from the great towns. The selection has been made on two assumptions, first that the preservation of *rare* species of fauna and flora cannot be made a prime consideration in the national parks of England and Wales. Secondly, that the proposed improvement in recreational facilities refers to characteristic physical recreations of the countryside, as rambling, climbing and boating, and to the intellectual recreation of nature study; and that it is not intended that national parks should extend or compete with the existing provisions for athletic games or indoor amusements.

The scenery of South Britain has great variety of natural character; no single area can be found which would be representative of the whole, and the minimum provision for preservation of natural character in national parks is therefore the selection of a supreme example of each principal type of landscape in those parts of the country which are not under standing crops or in fenced fields.

Excluding, therefore, the undulating lowlands with their fenced fields, let us examine those types of scenery in wilder state which are of recreative value from the beauty of their outlook, the refreshing qualities of air and climate and the active pastimes of the countryside. First is our coastline of 2300 miles, sought in the holiday seasons by numbers comparable to those who go to inland resorts. Of those who seek the seaside there is no inconsiderable part who desire the recreation of the shore itself and long to view the elemental ocean from a natural foreshore or unenclosed cliff. The opportunity has become much more difficult of recent years, largely owing to the facility which the private motor provides for the use of seaside residences for week-ends or holiday seasons.

Hence the immediate and urgent need for preserving in perpetuity the natural character of some portions of the coast and securing them for the recreation of people from all parts of the kingdom who retain the love of wild nature.

A systematic study of the coast soon reduces the choice of suitable areas for seaside parks to such narrow limits that the subsequent work of actual delimitation will be much simplified. It is in the west that the foundations of South Britain appear above the sea, the igneous rocks and hard ancient strata jutting in lofty headlands that withstand the waves, which pile up beaches in the sheltered coves between, and it is only at the southern extremity of the west coast that the shore lies open to the majestic swell of the Atlantic, which is one of the finest spectacles of our scenery. Here, too, in the terminal peninsulas of Cornwall and Pembroke, the proximity of the Atlantic brings an equable climate which permits the enjoyment of outdoor life on the very cliff in winter as well as summer, so that here the people may seek recreation throughout the year, an impossibility upon the fine stretches of the Yorkshire coast so attractive in summer time. A national park must at least provide a complete landscape, and the minimum for the Land's End park would be the cliff for 10 miles on either side of the headland; the maximum would be all between the suburbs of Penzance and St Ives. The area to be reserved need not, however, be large, only a mile or so inland, with the proper planning scheme for the adjacent interior.

The coast park of Pembrokeshire would be formed by the headlands of St Bride's Bay, terminating at St David's Head on the north and St Ann's Head on the south, the outlook from each diversified by off-lying islands, the haunt of sea-fowl. The intervening strip of coast with long accessible beach would be subject to a planning scheme which the proximity of a national park would render needful.

That the proposed coast parks in Cornwall and Pembroke are not adjacent to any large town is a suitable fact of situation for a park supported by national and not local funds. That they are in an extreme corner of the country is a disadvantage. Both peninsulas, however, already have a service of express trains capable of handling tourist traffic on a considerable scale.

I pass now to the consideration of the types of inland scenery which would be represented in a system of national parks, with the local examples which appear preeminent. We must begin in the west, for the history of British scenery began there, and we must follow it east-

Outlook from the Land's End

wards or, more exactly, south-eastwards according to the stages by which "Britain rose from out the azure main". As is the order of geological history, so is the character of the scenery. First in the north-west the rugged mountain peaks of the Lake District and the range of Snowdon, where volcanic protrusions of resistant rock are so carved as to leave upstanding peaks, which are the finest forms of natural scenery. Here are natural pastures which have long been a rambler's paradise, and crags, mainly of igneous formation, which provide the sport of rock climbing both in summer and winter. The residential parts of the Lake District could not be conveniently administered as a national park, but the upper levels (within a circle of about 12 miles' radius centred on Dunmail Raise, including some of the smaller lakes whose shores have not become essentially residential) are suited for the purpose. The Snowdon district of Carnarvonshire (with a small part overlapping the Merioneth border), having a length of 28 and an extreme breadth of 14 miles, could be almost all embraced in a national park, for the valleys with their lovely streams have not become residential to the same degree as the shores of the English lakes.

I pass from the scenery of mountain peaks to that of open moorlands with their sense of space. In the Devonian Peninsula we find in Dartmoor an upthrust of igneous rock, which forms a wild highland peaked by granite tors with stretches of open moor, unencumbered by the restriction of sporting rights, already established as a place of resort and recreation, standing greatly in need of measures of regulation, but which could be constituted as a national park with relatively small expense in the way of compensation. Nowhere in England is there a natural region of fine scenery which has a better defined natural boundary combined with a size sufficient for our purpose.

Passing from the eruptive rocks to those of the carboniferous period, we come to the great upland of the Pennine Chain and the Northumbrian moors, of carboniferous limestone, capped in places with millstone grit, having a length from north to south equal to that of Wales. This is the chief area of moorland in England. These moors are always splendid in their sense of space, and at the time of the summer holiday unequalled for colour and for fragrance of the bracing air. Freedom of access would be of the utmost value to the people of the great towns of Lancashire, Yorkshire and Tyneside, but grouse shooting constitutes a difficulty. Right of access would not make access safe during the summer holiday, and if rambling during the nesting season interfered with the birds and, later on, diminished the bag, there would be a serious economic loss both in letting value and indirectly in other ways.

It is the mountain limestone, hard as granite but soluble as chalk, which provides the pre-eminent examples of incised scenery, that kind of landscape in which it is the hollow, not the height, which dominates the eye. Here are gorges of architectural grandeur softened by vegetation clinging to the crags. In the dales of the Pennines and Northumbria are numerous examples, all naturally beautiful, some spoilt by ill-planned roads and bad buildings, some that require safeguarding under a town-planning scheme, and one, Dovedale, which among them all has probably a unique claim to be scheduled in the national park scheme, owing to the circumstance that its exquisite and untouched natural beauties are already the object of popular pilgrimage from the great industrial centres of both North and Midlands, and yet it is situated too far from any one of the great towns to bring it within the category of suburban areas where public parks should be maintained by municipal rather than by national funds. Dovedale is but a small part of the valley of the Dove, its length being about 3 or 6 miles according to different uses of the term, but it is the larger connotation which is appropriate to a national park scheme. The breadth to be included, in order to ensure an unspoilt view, is less, so that the area essential for preservation is much smaller than that suggested in the case of the Lake District, Snowdonia and Dartmoor. It is better, however, to leave it so than to endeavour to widen the area to conformity with the other districts, for the landscape of the surrounding uplands, although pleasing, is not pre-eminent, nor so attractive to the pilgrims of scenery as the heathery moorlands which are near by on the north.

The mountain limestone outcrops in the south-west of England also, and it is in that part of the River Wye which borders on the Forest of Dean that we find the most important of our river gorges. Government ownership of much woodland, mainly on the Gloucestershire side, greatly facilitates the preservation of the gorges of the Wye, and a sufficient area on the right, Monmouth bank, is included in the district proposed by Lord Bledisloe for the Forest of Dean national park.

I now leave the primary and harder rocks and come to the secondary, softer strata, which are for the most part in fenced fields, and in summer time in standing crops. Cupped between the Pennines and the hills of Wales is an agricultural Arcadia of which Worcester is a typical county. South and east is the band of lowland lias and the long upland of soft oolitic limestone continuous from the Dorset coast to the Humber, all under cultivation. In these parts there cluster the villages of old-world architecture in which English scenery is unrivalled. The desire for their preservation has led to the suggestion of a national park in the Cotswolds, where these characters are preeminent. It is, however, impossible to combine the requirements of preservation and of recreation upon a national scale in a limited agricultural area. It is true that townspeople can hardly do better than seek Arcadian England for scenic recreation, but they must spread and not concentrate, for where crops stand throughout the summer there can be no free rambling. The problems of preservation and enjoyment of scenery in these parts has more to do with the protection of the country resident than with any addition to the facilities of townspeople, which are already ample, so that as regards agricultural England the help of the Government is to be sought not in formation of national parks but in the direction clearly indicated in last year's resolution of the British Association and of the National Conference for Preservation of the Countryside, "urging His Majesty's Government to stimulate the employment by local authorities of the powers already conferred upon them by Parliament for the preservation of scenic amenity in town and country".

Continuing our progress towards the south-east corner of England we come to the distinctive scenery of the chalk, a rock so friable that its summits are rounded, so compact that its sides stand steep, yet soluble so that its hollows are shaped into perfect cups. This upland is consequently unrivalled for smooth continuity of billowy undulation, and is so conformable with the shape of cumulus cloud that the view on the Downs often presents a harmony of land and sky as perfect in its way as the harmony of sky and sea. Most of the extensive region of the chalk has been ploughed up, but here and there considerable unenclosed stretches remain of a natural turf in which the wild thyme grows, contributing to the fragrance which makes the air of the Downs delectable.

It seems desirable to include in the system of national parks an area of the chalk still in natural turf which shall satisfy the requirement of a natural boundary and have the advantage of affording an extensive and inspiring view. This combination occurs in Sussex, where rivers divide the range into well-defined blocks. The block at the back of Chichester is largely wooded; those at the back of Brighton and adjacent to Eastbourne specially concern these municipalities, so that as a park of open down for national purposes the 9-mile block between the valleys of the Arun and the Adur, from Amberley to Steyning, seems specially suitable. This area would not indeed be likely to prove an attraction for long sojourn by large numbers of adult holiday makers, and perhaps its quiet character would be impaired by such an influx, but the natural and occupational aspect of this kind of countryside would be of great value for the recreational instruction of the young during school holidays. This is an aspect of national parks which has so far received too little attention in England, but certainly constitutes no small part of their claim upon the national exchequer.

In Hampshire, upon later tertiary formation lying in the lap of the chalk, is the one national park that the country at present possesses—the New Forest. Here are not only the enclosed woods called Crown Coppices, but the "open" forest where commoners have grazing and other woodland rights. Here are preserved not merely the actual woodlands but also the economic life of Old England. The woodland glades, where the New Forest ponies wander at their will and the swine roam as in Saxon times, provide a scene which kindles the appreciation of natural conditions and stimulates the historic sense. Comprehensive planning for preservation of amenity in the towns and villages within and around the forest is all that is needed for the preservation and enjoyment of this national park, which is, indeed, all the more characteristically national that its administration is conducted partly by the Crown and partly by local representatives.

In the eastern counties we come to the latest-formed and lowest-lying of the geological strata, comprised under the names of "pleistocene and recent". Here before the eighteenth century were extensive areas of true fen, a country of mere and marsh and islet, where the economic life of the people was of a distinctive character, with harvesting of reeds, a fishing industry, and shooting of waterfowl. The district is now farmland growing heavy crops, a circumstance sufficient to prohibit the formation of a national park even if the spectacular horizon and sky view were sufficient attraction. But in East Anglia there still remains an amphibious region, half low-lying land with old-world villages, half reed-bound meres—the Broads, which are already a great holiday resort for boating. The heart of Broadland is a tract in Norfolk about 9 miles in length and somewhat less in breadth, traversed by the River Bure and its tributaries, the Ant and Thurne. The natural character and charm of the district can only be ensured by control of building within the area. The argument for including the district in the scheme of national parks is strengthened by the fact that in reference to the recreation of the people the claims of the boating fraternity should not be ignored. Moreover, in this district the preservation of the natural banks or shore can be secured at a small fraction of the cost of compensation in a district such as the Chiltern Reach of the Thames. But the value to the nation of a conserved and protected Broadland is not confined to its contribution to the physical and mental recreation of adults, but largely lies in the lessons to be learnt by holiday classes of the young from this remarkable survival of a peculiar type of scenery and country life.

I have now considered each of the principal types of scenery in South Britain in relation to the establishment of national parks, or alternatively to the preservation and enjoyment of scenery by other means.

I have finally to examine the situation of the suggested sites in relation to the principal industrial areas, and for this purpose it is convenient to choose representative towns as centres of circles of 120-mile radius. These representative towns are London, Cardiff, Birmingham and Manchester with Leeds. If we add Glasgow and take note of the districts, which, as I learn, have been suggested for national parks in Scotland, we shall see at once that, although in reference to allocation of funds from the Exchequer for national parks in Britain the island is one and indivisible, yet in reference to accessibility by the mass of the people the proposed parks of Scotland will be outside the 120-mile radius from the urban centres in England and Wales, and of the proposed parks in South Britain only that in the Lake District is within 120 miles of Glasgow. Indeed, most of the propositions for national parks in Scotland relate to the northern part of the Highlands, within 120 miles of Glasgow but 250 miles from Manchester and 400 from London. Of the suggested inland parks in England and Wales the Snowdon Range is within or on the 120-mile circuit from Cardiff, Birmingham, Manchester and Leeds; the Lake District within this circuit of Manchester, Leeds and also Newcastle-upon-Tyne; Dovedale within the circuit of Leeds, Manchester and Birmingham; the Forest of Dean and, likewise, the New Forest, within or on the circuit of London, Birmingham and Cardiff; Dartmoor within that of Cardiff and

Bristol, and the Sussex Downs and Norfolk Broads within that of London.

Thus the suggested sites are more evenly distributed with reference to the urban areas than might at first seem possible, having regard to the essentially unsymmetrical distribution of the physical features of the country.

The above proposals are made on the assumption that the arrangements as to lodging, camping and motoring in national parks will be regulated in each case according to the local features, and so as not to impair natural character or destroy the sense of quietude.

2. NATIONAL PARKS IN WALES

(Evidence submitted to the National Park Committee on behalf of the Council for the Preservation of Rural Wales, Feb. 4th, 1930)

In the preparation of our scheme for national parks in Wales, we have noted the fact that the word "national" in the terms of reference to the National Parks Committee is employed in the sense of British. In this connection we may remark that the position of the unenclosed and wild country accessible from most of the industrial centres of England indicates that a considerable part of the area to be devoted to national parks in South Britain should be situate in Wales. In our selection from the many areas of sufficiently wild and open character we have taken account of those historical associations which will enhance the recreative value of the parks for the Welsh people, and, we would add, educative value to the English visitors. Thus the proposed park in Snowdonia includes the mountain stronghold of Gwynedd, and that on the coast of

Pembroke the old ecclesiastical capital, St David's. Apart from historical associations these areas are second to none in Britain for the purpose of national parks on account of their natural characteristics.

THE MOUNTAIN PARK IN SNOWDONIA

The panorama of Snowdonia from the direction of Anglesey provides the best example in Britain of a true range of mountain peaks. These are, moreover, the mountains most accessible from London, the industrial centres of the Midlands, Liverpool and South Lancashire. The four masses are Moel Hebog, Snowdon proper, the Glyders and Carnedds. Parallel with this main range is the shorter range on the south-east culminating in Cnicht and Siabod. This subsidiary range completes the enclosure of the beautiful valley connecting Bettws-y-Coed with Portmadoc by way of Capel Curig, Pen-y-Gwryd and Beddgelert. From Capel Curig, Pen-y-Gwryd and Beddgelert the whole breadth of the main range is traversed by low passes provided with motoring roads. The mountainous area is of lenticular shape with a length of about 24 miles and a breadth of about 12 miles, and possesses in an unusual degree the well-defined natural frontier without which a national park would introduce an element of artificiality tending to counteract its purpose of preserving natural character. The region, which is already sufficiently provided with motor roads, is in its recesses a paradise for the pedestrian. It provides rock climbing in both summer and winter, a recreation which would be facilitated by the establishment of a few climbers' huts on the Alpine Club model. The unfenced sheep pastures afford ample scope for rambling; shooting rights do not cause such difficulties as those encountered on Pennine grouse moors or Scottish deer forests, and few regulations would be needed except for the prevention of litter. Neither are the charms of rambling merely those of wild, open spaces, for the natural rock-gardens are of exquisite beauty, and brimming cascades leaping down the mountain are compensation for the rainy climate. The upper courses of the rivers would be within the park proper; the lower reaches, winding in pools and shallows between wooded banks, traverse inhabited areas which would be dealt with under the planning scheme which is a necessary accompaniment of the formation of a national park. In the inhabited and cultivated parts, within the enclosing perimeter of the park, it would be essential that not only the sanitation and stability of new buildings, but also their form, colour and texture should be subject to regulation. The boundary between the parked and planned areas can only be drawn after a field survey by experts and local residents.

We have dealt at some length upon the characteristics of this district, because we believe that, even if considerations of national finance should make it necessary that the scheme of national parks in Britain should come into operation gradually, the mountain park of the Snowdon district should be comprised in the first selection of inland areas on account of its pre-eminent natural beauty, of the historic interest of the great castles which guard the entrances, of the economic advantages consequent upon the comparatively small sporting value of the mountain pastures, and on account of its geographical relation to urban areas, which is of a kind suitable for a national as distinguished from a municipal park.

THE COAST PARK IN PEMBROKESHIRE

The natural scenery of the coast culminates in cliffs, as that of inland scenery in mountain peaks. The vast expanse and distant horizon of the sea, viewed from the lofty headland, is unsurpassed in elemental grandeur; in the sheltered cove, where waves break on a beach unspoilt by artificial structures, the scenery is more purely natural than on mountain pastures. The preservation of the natural character of the cliffs and their foreshore requires more elaborate measures of legislation and administration than are needed for mountain summits, because the latter are unsuitable for residence, whereas sites for villas and bungalows are sought close to the cliff with gardens reaching to the very verge, thus preventing public access to a priceless view. The Council is of opinion that the coastline of Pembroke from Strumble Head to St David's Head, and thereon to a point opposite Caldy Island (thus including St Bride's Bay and the outer part of Milford Haven), should be dealt with on principles somewhat similar to those suggested for the Snowdon area, the cliff parts, now in their natural state, being made a national park, with the islands as its bird sanctuary, and the low-lying parts of the shore and its villages made subject to regulations relating to scenic amenity under the Town Planning Acts. The seaward boundary of the district is estimated roughly at about 50 miles. We are aware that this splendid stretch of shore is somewhat remote, but a review of the whole coast of South Britain and its watering places reveals the fact that there is very little choice of areas for coastal parks which have a sufficient length of open cliff combined with a climate favourable to outdoor recreation in winter as well as summer. The essential measure of preservation of the cliff portion is the placing of the building line at a sufficient distance from the cliff. This line would be the boundary of the park, but the area at the back would have to be made subject to a planning scheme to secure the amenity of buildings. The enhancement of value in this back block, owing to the preservation of a broad pleasance in front, would be great. The breadth of the strip on which building would be prohibited can only be determined by survey on the ground, conducted by an expert in planning in conjunction with the landowners and representatives of local authorities.

The project of establishing national parks upon the coast is still somewhat novel. Their institution may entail prolonged negotiations, but the benefit which would ultimately result from them can hardly be doubted by any person of taste who has observed the rapid sophistication and even vulgarisation of our shores.

CHAPTER II

TREES AND BUILDINGS IN RURAL ENGLAND

1. TREES IN ENGLISH LANDSCAPE

(*From* The Tree Lover, *no. III, April* 1933)

Englishmen have an unusually keen perception of the beauty of trees. The charm of our well-wooded countryside is no mere accident of soil and climate, for the neighbouring countries of Europe belong to the same botanic region. When we land at Dover or Folkestone on our return from a visit to the continent, and, taking the special train for London, travel swiftly and smoothly through the county of Kent, we note with a new clearness of vision how the pattern of an undulating landscape is improved when trees are planted so that they may spread their boughs, and choice specimens are left standing after the passing of their economic prime, so that they shall display the beauties of venerable age.

Trees have certain qualities of beauty which the average man sees with understanding, and others which he can recognise as soon as they are pointed out, deriving thereafter greater enjoyment than that experienced while these visual qualities influenced only the subconscious mind.

The foundations of aesthetic appreciation lie deep down and have as yet been only partially uncovered, but I think it may be said with confidence that the strong impression of the personality of the tree is related to the fact that it stands upright, as we do, and with trunk and limbs. That some similarity of form is no mere fancy but a visual fact is strongly suggested by the observation that human figures appearing in the woodland scene have a special charm which is lacking when they are viewed in the open country.

Particular attributes are associated with each of our favourite kinds of tree according to the peculiarities of form. We think of the rugged endurance of the oak, with its gnarled stem and arms squarely bent as if elbowing an antagonist aside; of the gentle suaveness of the birch yielding readily but regaining the purpose and direction of its growth with graceful curves of pleated stem; of the pine, erect and stately as some tall sentinel.

Another aspect of the personality of trees is their witness to the conditions of environment, as the thorn which stands with curling, twisted branches on a windy brow of Dartmoor, or the elm with trunk bending in an arch across a country lane, where the rush of the sea wind sweeps over the Purbeck Hills.

In other environments it is not by their posture but by the apparently conscious selection of a site that the tree stands as a witness to natural conditions. This attractive characteristic is displayed in the luxuriant climate of the Lake District, where a little soil, lodging in a crevice of the rock, suffices for the growth of a rowan tree, with roots which grapple the bare surface of the crag.

More subtle, but not less important than these functional relations in their contribution to scenic beauty, are the purely visual harmonies of the form of trees with that of the surface of the land. The rounded hills and hollows, typical of the English Lowland, are conformably patterned by the billowy, mammiform surface of the woods and coppices; and bushy, quickset fences complete the pattern of gentle curves.

The combination of ancient, spreading trees spaced far apart upon smooth sward, which we know as the scenery of the English park, occurs in Nature as the tropical savannah, and it is a tribute to the genius of our ancestors for artistic forestry that the beauty of the tropical savannah is usually characterised by reference to the English park, as though the latter were the prototype.

Where the hard rock appears and the hills or mountains culminate in pyramidal peaks, the pointed larch emphasises the features of the land by an accessory pattern of harmonious form.

The pictorial effect of a tree as a *repoussoir*, or feature enhancing the apparent distance of objects in the background, is noticed by us for the first time when travelling abroad, for instance in the lower Rhone valley when on the way to the Riviera, or in crossing the plain of Lombardy on the return journey by way of the Simplon or St Gothard. In travelling from east to west across the southern counties of England the effect becomes noticeable when we reach the valley of the Severn and look across to the mountainous forms of the Malvern range. The Lombardy poplar, which excels in this effect, has also especial picturesqueness on the broad stretches of perfectly flat meadows which border some of the principal tribu-

The Lofty Elm

Sandhurst, Berks

taries of the lower Severn, for a combination of horizontal and vertical lines always leaps to the eye. In the Fen country this effect of slender trees is less noticeable because the tall church towers usurp their function. The trees dotted over the countryside in this part of England contribute to the appearance of great distance, not so much in the manner noted in plains bordered by mountains, but by the etherial tone of the tree-tops seen above the nearer skyline of the land.

In our climate, which is subject to the moderating influence of the ocean, the four scenic seasons of the year are of nearly equal length, the seasons of the unfolding and the fading leaf being both prolonged, a contrast to the sudden rush of spring and the quick onset of winter in continental climates of the same latitude. The English spring is perhaps less spectacular, but the phases of the foliage can be better observed; the change from the solid, shining bud to the young leaf, opening as the wings of a butterfly, in tender green, translucent to the rays of the returning sun; and the strengthening chord of colour as the foliage of the early varieties of tree deepens in tone and each later kind sustains in succession the lighter tint.

The delights of spring are largely due to a co-operation of the senses; for the song of birds and the fragrance of leaf and flower not only give pleasure in themselves but so quicken the aesthetic side of Man's personality that the enjoyment of visual beauty is enhanced. The beauty of autumn has less of this adventitious aid. In visual character these seasons are distinguished respectively by multitudinous detail and broad effect. In coppice woods this contrast is especially noticeable. In springtime the unequalled attraction of the wild flowers makes even the canopy of translucent leaves subordinate, but when we walk in the woods on a still, sunny day in mid-autumn with a leafy canopy of green and gold and bronze arching over a floor of russet bracken, the whole surrounding sphere is so spread with rich and joyous colour that the satisfaction of the visual sense is of itself sufficient to transport the mind to those realms of bliss where time is no more and not only sorrow but sin itself seems to pass away and leave us in the state of Perfect Being.

When the leaves are stripped, trees are dependent upon soft, mellow sunshine for their beauty by day, and in the evening hour on the banded colour of a clear sunset sky. Anyone who has seen a tropical forest stripped of leaves by drought, the whitened boughs like the bones of a skeleton bleaching in the pitiless sun, will realise how tenderly the misty sunshine of our open winter treats the lattice-work of oak and elm, and the grey lichen which coats the trunks and boughs.

The thick hoar frost, which in our moist climate comes with the first nip of cold, forms best on the most slender shoots, and so in frosty sunshine the articulated tracery of the birch is the culminating beauty of the woodland scene, the silver filigree perfectly displayed by a background of blue sky.

Here and there in the southern and midland counties where hills rise in isolated eminences capped with coarse gravel, notably that of the bunter sandstone, a clump of weathered pines provides a perfect focal feature; but amidst the broad-leaved forest of the English plain, pine plantations spoil the pattern, and in that most precious relic of Old England, the New Forest, no more pine planting should be allowed.

Yet the single pine of venerable age, here and there, in the English meadowland, plays a pictorial part, for in the brightest days of summer it carries the shadows of foliage down to greater depths. Neither should we forget the effect of the broad brushes of the pine tree's foliage in the winter twilight, when, after the fading of the orange glow in the west, the slender, bare boughs of the deciduous trees are no longer conspicuous. Then the bolder silhouette of the pine tells magnificently against that calm tint of twilight which intervenes between the cobalt sky of day and the deep indigo of night.

Some of the influences of tree-lattices upon the functioning of the eye are very elusive although pictorially important. When the sun shines down amidst the maze of a wood with coppice undergrowth, there is no feeling of being walled-in, although the distance visible is small. Were it not for a satisfying sense of space the scene would lose much of its charm. The third dimension in scenery, that of recession, is frequently recognised by inference only, as from relativity of tone, but is more impressive when stereoscopically perceived. The three-dimensional lattice of the trees provides the very means for bringing this faculty of the eye into use, making space itself, so to speak, stereoscopic, giving it "solidity" in the geometrical, as distinguished from the physical, sense. The stereoscopic action of tree-lattice is also an important factor in the scenery of cities, being largely responsible for the enhancement of architecture by the boulevard method of planting, an effect best achieved when there is a central, pedestrian avenue which permits the buildings on either side to be viewed through the trees. The result is an increase of apparent size, and an enrichment of atmospheric tint.

2. THE PICTORIAL UNITY OF ARCHITECTURE AND NATURE IN ENGLISH SCENERY

(*National Conference for the Preservation of the Countryside, Ambleside, Oct. 13th, 1929*)

We are still backward in the critical study of the pictorial unity of architecture and nature in rural scenery, although this harmony is an important motive of the landscape painter. The scenic value of rural architecture is two-fold, the harmony of its parts and the harmony of the whole with the natural surroundings. The latter harmony originated for the most part accidentally, but having once been achieved is easily recognised and should be purposefully continued. The spire of the country church has greater charm when seen amidst tall elms, for it embodies in perfection the character of tapering height towards which the trunks and limbs of the trees approximate. In the Fen district church towers catch and emphasise the rectangular pattern of the landscape. In the Weald country and Worcestershire cup-shaped hollows in the hills provide a background which displays the farmstead and cottage to great advantage, and the winding valleys where streams break out beneath the Berkshire Downs are a perfect setting for the church and hamlet by which they are

The Mermaid Inn

Rye, Sussex

adorned. In the English Lake District the rough local stone employed in the old peasant architecture is strikingly suited to the rugged crags. The old farmsteads and cottages of the midland and southern counties, admired by persons of taste throughout the world, are also of local material; the later building, which is so often a blot on the landscape, is frequently in material brought from a distance. These facts are commonly cited in support of the theory of a natural pictorial affinity between local building material and the landscape of the district, but in the undulating and wooded country of the midland and southern counties the disharmony of the nineteenth-century building is not really due to physical but historic causes. Mechanical transport came at the same time as mechanical construction, so that the employment of material obtained from a distance began at the same time that mechanical rigidity of line and mechanical repetition of form succeeded the artistic freedom of skilled craftsmanship. At about the same time originated that raw tint of red brick and tile which was so disastrous in a verdant landscape. It is fortunately no longer necessary to wait for the growth of creepers to subdue the tone of staring walls, for under the guidance of architects of the modern school, villas and cottages are being erected in brick and tile of a tone which takes its place quietly in the landscape, and the lines of roof and gable in these twentieth-century buildings are relieved from rigidity not by the limitations of handicraft but by the touch of artistry.

The dominant types of good village architecture in England were laid down in Tudor times, an era of extensive building. Geographically they are mostly comprised in the long band of oolitic limestone from Dorset to Lincoln, in the country in the south and east, in Devon, and in the districts which lie between the oolite and the hard rock of Wales and the Pennines. The colour of wall and roof of a Cotswold cottage ranges from lichen grey to straw tint. In April sunshine the villages of the Cotswold valleys form with their sylvan background a pattern of these colours, for the pale grey trunks and boughs of elms, the lichened gates and fences, the sere grasses of field and hedgerow, the yellow-green catkins of the copses are integrated by the massed grey and buff of church and cottage and stone-roofed barn.

Late in May when the leaves are out and meadows of a uniform bright green, the pattern of vegetation in the Weald of Kent is of dark objects upon a light background, the dark parts being the shadowed sides of trunk and bough and leaf. The gabled ends of old wooden barns, black painted under reddish-brown tiles, carry this pattern on to its consummation, and therefore at this season do more than the fine old red brick farmhouse to enhance the beauty of the natural landscape.

CHAPTER III

THE SCENIC AMENITY OF GREAT BRITAIN

(*From* Geography, the Quarterly Journal of the Geographical Association, *Sept.* 1934)

INTRODUCTION

The passing of the Town and Country Planning Act (1932) has given Local Authorities in Great Britain power to take measures for preservation of scenic amenity, but it is difficult for the member of a Local Authority to think clearly and consecutively upon the conditions which determine scenic amenity, because his mind is subjected to a harassing oscillation between the question of what is desirable and what is practicable.

Here the modern geographer can help, for "scenic geography", the aesthetic aspect of our science, has for some years past been systematically studied.

In order to draw the line between the scenic and other amenities with which the Town and Country Planning Act is concerned, it is well to recall the fact that the word "scenery" is derived from the Greek *skēnē*, a tent or stage, and that "the natural scene" may be properly regarded as the stage on which Nature plays the drama of the senses. Thus, not only the form and colour of the landscape, but the song of birds, the sounds of running water, the fragrance of the field, and the mere breath of pure, fresh air are scenic amenities. Discordant noises are an offence against one of the amenities, smoke doubly offensive, since it disfigures buildings and contaminates the air. On the other hand, such matters of planning as the supply of pure water, or the interior accommodation of houses, do not come within the scope of scenic amenity.

The geographical unit which we have to study is the area to which the Town and Country Planning Act applies, namely, the whole of Great Britain, Northern Ireland not coming within its scope. Consequently, the region is clearly defined as concerns natural features; a single island, the off-lying islands being so close and relatively small that they do not impair the essential unity. In respect of architectural features and historic scenes, however, the matter is not so simple. Although as regards foreign affairs the people of this island are one and indivisible, a British nationality, as was made manifest in the crisis of August 1914, yet the island contains three Home Countries. Looking more closely into the matter, we see that the natural features of England and Wales are dovetailed together by the valleys of the Usk, the Wye, the Severn, and the Dee. The mountains of Snowdonia are readily accessible to the industrial population of Midland and North-western England. The Scottish Highlands, with their special beauty of lochs and glens, are, on the contrary, cut off from South Britain, first by the Southern Uplands, where the natural features resemble those of the English northern counties, and, secondly, by the urban belt of the Scottish Lowlands. A convenient resort for the great population of the Clyde area, but distant from

South Britain, it may seem at first as if the Highlands of Scotland were too remote for their scenery to be important in the national life of England and Wales. But we must look ahead and make due allowance for the fact that the increase of touring consequent upon the introduction of the motor and aeroplane has not nearly reached its limit. Moreover, the Highlands have long been a resort for the wealthiest Englishmen, and we have only to recall the history of the playgrounds of Europe to see that the class unkindly called "the idle rich" are very diligent as explorers of pleasure resorts, and that the great mass of holiday-seekers presently follows in their footsteps.

It is sometimes said that the great spaces of the Highlands need no safeguarding; but let us take warning from the alternative plea for *laissez-faire* which follows immediately upon encroachment—namely, the protest that "it is too late to do anything". There is also an economic reason for considering the landscapes of England, Wales, and Scotland as one scenic region. The scenery of Great Britain is exercising an ever-widening appeal overseas, and there is a clear indication that if the scenic amenity of town and country be carefully guarded, our island will take a leading, perhaps the first, place among the touring resorts of the world.

TOWN AND SUBURB

The present movement for guarding our rural scenery was induced by the spate of building which followed the cessation of house construction during the Great War, by the disturbance of rural peace, due to increase of motor traffic, and the disfigurement of villages by poster advertisement. In order to combat these evils a number of societies concerned with the preservation of particular amenities united to form a Council for the Preservation of Rural England. The success of this organisation soon led to the formation of a Council for the Preservation of Rural Wales, and an Association for the Preservation of Rural Scotland. But if we are to envisage correctly the relation of scenic amenity to national life, we must not confine our attention to rural Britain. The recent encroachment upon the countryside is the second great inroad upon the scenic amenity of the island. The first, which dated from about the beginning of the nineteenth century, was upon the beauty of the towns. The lay-out of their multiplying streets was unplanned, and in the "elevations" of their new mechanical structures no educated guidance replaced the vanished hand of the craftsman. Unfortunately, no Council for the Preservation of Urban England was then formed.

In the course of the nineteenth century, picturesque market towns and beautiful cathedral cities were embedded in a maze of unsightly streets, whilst in the coalfields of South Wales, the Midlands and North of England, and the Lowlands of Scotland, there grew up great industrial towns where art was ignored and from which Nature was expelled. Here man created a desert and called it a city. In the early days of the industrial revolution there were the beginnings of an urban aristocracy in Britain, but smoke and hideous surroundings broke up these social groups and thereby impoverished the cultural development of the nation. From the earliest days which any of us can remember there has been a steady exodus of the well-to-do, who retained an office in the town but made their home elsewhere.

We have still, however, a few examples of urban scenery from which Nature has not been expelled. Of these the most important is the West End of London, the seat of the Royal Court, the social centre not only of a nation but of an Empire. Next, on a much smaller scale, are the collegiate parts of Oxford and Cambridge, and lastly the precincts of the English cathedrals, where the triumphs of Gothic architecture are enhanced by a verdant setting in a manner lacking to the cathedrals of the continent.

Reconstruction in towns to meet new requirements involves many tasks of protection from vandalism, and it is much to be hoped that the sentiment which protects ancient monuments may include the dignified architecture which is a precious heritage from the eighteenth century. As regards the close masses of residential streets of Victorian times in London and other great cities, no such consideration need interfere with the rebuilding which their want of modern conveniences demands. Now that steel instead of stone or brick is the structural material of towns, the present population can be accommodated upon one-third of the ground space by increasing the height of the houses, with no diminution of sunlight, and almost all the added space will be available for town gardening and afforestation.

Meanwhile, the progress of invention and the need to

minimise service in the home are reducing the smoke of domestic fires which hitherto has hindered the growth of vegetation in the towns and also militated against the alfresco habit which adds so much to the cheerfulness of life in cities. Smoke not only besmirches clothes, but absorbs both the thermal and actinic rays of the sun. If, however, the atmosphere of our paved cities were smokeless, their climate would be actually better for sedentary recreation than that of the countryside, on account of the protection from the exhalations of damp ground. It is damp ground more than winter cold which has hindered us from acquiring the habit of taking our rest and meals out-of-doors.

When winter has passed, the unfolding of the young leaf in the parks and boulevards of a smokeless city would impart the scents of the country to the atmosphere of the town and satisfy a craving which is natural to us all, for the air of a desert is alien to our race, born to breathe the fragrance of the field in a green and pleasant land.

The conversion of every city into a garden city is, moreover, the most important pictorial consideration in the replanning of our towns. The eye becomes fatigued by an environment of formal lines, harsh texture, and colours mostly cold or drab. Therefore, even though the architecture of a city be admirable, yet it requires the softening influence of a verdant setting.

The repulsive effect of life in a desert of bricks and mortar has created radial suburbs following the main roads, arms of an octopus whose body is the town. By this mode of growth the benefit of fresh air is bought at an

White Mill Bridge

Near Stourminster Marshall, Dorset

unnecessary price. The social result of the lay-out is very dull, for the inhabitants of the radial suburb have neither the close and varied intercourse of the town nor the seasonal round of occupation which enlivens an agricultural community. The actual scenery of such suburbs is correspondingly lacking in interest owing to the circumstance that it illustrates but a fraction of the life of its people. The actual pattern of the scene is also defective. In a well-planned city the architecture is displayed against a verdant setting; in the country any good architectural feature forms a suitable focus for a landscape of informal lines; but in the unplanned suburb the confusion of these forms produces a state of camouflage, which is the bane of beauty.

The scenic advantage of preserving a green belt round every town and developing compact "satellite", instead of radial, suburbs has obtained such general recognition as to ensure that public opinion will support those Local Authorities who grapple courageously with the administrative difficulties.

It is very desirable to preserve as far as possible the natural banks of our winding rivers, but of yet greater importance to scenic amenity is the planning of the foreshore of our seaside towns. It is since the beginning of the nineteenth century that these have become the great pleasure resorts of the nation, and consequently their architecture has very little of the beauty of earlier inland spas, as Bath, which date from a better period of architecture. The actual eyesore, however, is the jumble of bad structures upon the foreshore; the piers, kiosks, huts, bungalows, and all kinds of abominations which desecrate the prospect of the sea. Fortunately, most of them are not of a very enduring character, but if their redesign is to be satisfactory, two conditions will have to be satisfied—first, the design must be entrusted to professional architects, and secondly, the members of the profession must turn their attention to the special task of designing for the background of the beach and sea.

COUNTRYSIDE AND COAST

Those who seek solitude by the sea have to walk farther than the visitor to a holiday resort among the moors, whence pedestrians disperse in all directions. Unfortunately, the pathway by the shore is often interrupted by private enclosures reaching to the edge of the cliff or the high-water mark; and it is time that further encroachments of this kind should be prohibited. Yet even when this protection of the public has been secured, something more will be needed to provide access under conditions of peaceful harmony to the coastline, which is the special scenic heritage of an island people. In order to achieve this, some considerable stretches of the shore should be preserved in their primitive wildness as national parks. In selecting sites for these we have to seek places which combine grandeur of cliff scenery with advantages of winter climate. In the north-western Highlands the winter day is too short for the holiday-seeker; on the Cleveland shore of Yorkshire the east winds of spring are too cold. Other districts are too near the larger watering places for our purpose, and so, by the process of exclusion, we find that two counties emerge pre-eminent—Cornwall and Pembroke—the peninsulas which flank the entrance to the Bristol Channel. No time should be lost in securing adequate stretches of cliff scenery in these Atlantic outposts, where the winter climate is mild and sunny.

Starting from London and traversing rural England by any one of the radiating roads or railways, the dominant character of the scenery is agricultural, a country of fields, divided by quickset fences with hedgerow timber, and well wooded with deciduous trees of spreading growth, which adorn the undulating surface with a smaller pattern of conformable shapes. The green fertility of the whole enhances the restful effect of the suave and easy lines, but were this the whole picture, the scene, although quietly pleasing, would be wanting in emphasis from lack of focal features of bold outline. Such features, however, in which Nature here is lacking, are provided by the architecture of cottage, homestead, and hamlet, and of the village street with its crowning feature—the Gothic church. The combination of the English village, with the setting of field and hedgerow and coppice, is an Arcadian scene unrivalled elsewhere in Great Britain and unsurpassed in any part of the world. Moreover, the great variety of geological formations has resulted in a very interesting variety of local styles of building.

There is, at the present time, no greater obligation in the matter of scenic planning than the protection of the English village. It must be guarded against incorporation in "ribbon development", maintained in architectural unity, and secured against the disfigurement of cottages by posters. Moreover, the quietude, which before the advent of motors made the village street a place of pleasant loitering and friendly gossip, must be maintained, those situate on main roads being provided with a by-pass.

Let us now extend our tour into Wales, following the connecting valleys. Here the market towns lack the charm which still lingers in those of the English border counties; and the village streets are not so beautiful as those of England,[1] but the natural features of the landscape are on a grander scale. Bold heights rise above the valley, where the stream flows with a stronger current; and, on the great spaces of unenclosed upland, the sweet, fresh air is invigorated by the aromatic fragrance of wild thyme. In North Wales, hard, ancient rocks crown the landscape with pyramidal peaks; and Snowdonia, viewed from the west, presents the spectacle, unequalled elsewhere in Britain, of a true mountain range.

It is in Wales that the Anglo-Norman castle survives in its most impressive form. Magnificent monuments of fortress-habitation guard the gateways of Snowdonia. This noble country (still called by the name officially

[1] The detached houses are, however, often picturesque. Professor Fleure sends me the following note on this subject: "The farmhouse and cottage architecture of Wales has taken on forms that respond to various regional environments. The beams of the great oak, *Quercus robur pedunculatus*, furnish material for 'black and white' houses in east Montgomeryshire and east Breconshire. To the west, where highland and sea winds cut out the great oak, we find houses in Merionethshire and parts of Carnarvonshire picturesquely built of blocks of solemn igneous rock, while in Cardiganshire many a cottage is of clay blocks, with the clay bound together by straw and often coloured white or pink. The Builth area, with its lava blocks, and Pembrokeshire with its so-called Flemish chimneys, give other variants."

The Pine Wood by the Sea

bestowed at the time when Latin was the language of our statutes) is at once a Welsh Valhalla and one of Nature's shrines for the English pilgrim of scenery. Snowdonia should undoubtedly be protected by a dual planning scheme. The open uplands should be constituted a national park, which need not involve expropriation, and the surroundings made subject to those careful provisions with regard to lay-out and elevation of buildings which will be required in the neighbourhood of an area reserved as a national resort.

The natural features of Great Britain are felt to be in great measure the common heritage of all natives of the island, but the historic sites and monuments of England, Wales, and Scotland have for the most part a separate appeal to the three nationalities. There is, however, one outstanding exception—namely, Hadrian's Wall, which still stands as a conspicuous feature of the landscape for miles of upland moor where Britain narrows at its central isthmus. This historic monument is equally the concern of the Southern Kingdom, formerly a province of Mediterranean civilisation, and of the Northern Kingdom, which looks back with pride to the barbaric independence of Caledonia. Few can gaze unmoved upon the scene at the camp of Housesteads, the Roman *Borcovicium*, where the pavements are scored deep with the ruts of the chariot wheels, and away to the north the limitless prospect is "stern and wild" as in the days of the Pictish tribes.

The preservation of ancient monuments is, however, not in itself sufficient to maintain the amenity of historic scenes, for no feature is truly picturesque unless seen in a suitable setting. Something has already been done in this matter under a recent Act, which has enabled the Office of Works to protect part of the immediate environment of Housesteads, but the position will not be satisfactory until an area of moorland some 9 miles by 5, centred in the camp, is permanently preserved as a national park, and the neighbourhood as far as Chollerford and Gilsland protected from vandalism by a scheme under the Town and Country Planning Act.

Reserving the Lake District to the last, as the epitome of British scenery, I now pass on to Scotland. Here, as in Wales, we no longer find so perfect an architectural focus as the English village, neither does the Gothic cathedral survive in its beauty, nor the cathedral close. But a new feature of castellated architecture appears, which has a singular attraction as the crown of the natural landscape. The ancient volcanoes of Scotland have long since been stripped of the cinder cones, but basalt blocks and necks of consolidated lava have been left standing above the Lowland plain. Here castles, as those of Edinburgh and Stirling, crown the volcanic crags and carry defensible precipitancy to a culmination towards which Nature seems to strive but never fully to attain.

There is, moreover, an architectural feature of Scottish castles which enriches the scenery of Great Britain with an historical association of particular interest, the conical turret, a Gallic touch notably absent in England and Wales.[1]

[1] Professor Fleure writes: "There is many another French touch in the Scottish burghs. Visitors to Stirling know well an old mansion near the

The perpetuation of local style in architecture is important for the decoration of the rural landscape as well as for historical association, since *pattern* is the foundation of scenic beauty. In the countryside it would not be economically possible, even if desirable, to pull down and rebuild the houses; and so, unless the new buildings conform in style, the architectural pattern of the landscape will be camouflaged. Such is the effect in some parts of the Lowlands, where the stucco wall and red-tiled roof, seemingly borrowed from the environs of London, are seen side by side with the stern solidity of the old stone house.

The districts of the Highlands usually suggested for reservation are mountainous groups, as the Cairngorms, suitable as national parks for rambling and climbing; and regions such as can be found in the north-west, which, from their remoteness, are suitable for Nature reserves, where rare animal species can be protected. There is, however, a special obligation to introduce at an early date a comprehensive planning scheme for one or more of the Highland glens—as, e.g. Glen Affric. For this there are two reasons, one general and one local. The general consideration, which has to do with Great Britain as a whole, is the circumstance that valleys closely encompassed by steep hills tend to become avenues of traffic and frequently areas of congested building. In England this process has gone too far for much to be done in the way of preserva-

church and castle; it might have come from some town in the Paris basin, ready made. At the same time, the stepped gables of some old houses in Stirling and elsewhere in East Scotland remind one of links with Flanders and the Hanseatic area."

tion. Of the two other constituent countries, the chief examples of glen scenery are in Scotland. The Highlands are mainly a furrowed plateau, not a region of bold upstanding forms, and the deep glens, with their lochs and the dark corries, are the most characteristic feature of the landscape.

I come finally to the consideration of the amenities of the English Lake District, where the Fell country, which forms the nucleus of the area, is second only to the New Forest in its claim to be treated as a national park. The scenery of the English Lake District, within an area of suitable size for a national park, combines, in a manner not met with elsewhere in Great Britain, features specially characteristic of the three national divisions of the island. It unites the beauty of the Scottish lochs with the mountain grandeur of North Wales, whilst the well-wooded valleys and their rustic architecture have the Arcadian character of Old England.

CHAPTER IV

THE SCENIC AMENITY OF SOUTH-EASTERN ENGLAND

PRESIDENTIAL ADDRESS TO THE REGIONAL SURVEY SECTION OF THE SOUTH-EASTERN UNION OF SCIENTIFIC SOCIETIES

(*From* South-Eastern Naturalist and Antiquary, 1935, *vol.* XL)

INTRODUCTION

The region which I have chosen for the study of scenic amenity at this conference is the geological province which comprises the counties from Norfolk to Hampshire where the constituent societies of the South-Eastern Union are situated, as well as those parts of Wiltshire and Dorset which are adjacent to Hampshire, the county in which our present annual meeting is held.

This is the natural region comprised by the chalk and folds of the chalk, a triangular area of compact and symmetrical shape bounded on the north-west by the escarpment of the chalk from Dorset to Norfolk, on the east by the North Sea and on the south by the English Channel. The escarpment is marked by the Dorset Heights, Salisbury Plain and the Marlborough Downs in Wiltshire; the White Horse Hills in Berkshire; the Chilterns, the Dunstable, Luton and Royston Downs and the Gog Magog Hills in Oxfordshire, Buckinghamshire, Hertfordshire, Bedfordshire and Cambridgeshire. After this the chalk sinks lower and is overlaid in Suffolk and Norfolk by glacial and other deposits. It is, however, again displayed in the cliffs at Hunstanton at the north-western corner of the Norfolk coast, 200 miles north-east-by-north of Lulworth. Viewing England as a whole, and taking account of areas, it will not be inapt to describe the corner cut off by the continuous line of chalk as the south-eastern quadrant of England.

Except in the East Anglian plain, the solid modelling of the south-eastern quadrant of England is determined by the folds of the chalk. The London Basin, which includes the consecutive valleys of the Kennet and the Lower Thames (not the valley of the Upper Thames above Goring Gap) is filled with tertiary gravels and clays which lie on the top of a syncline or downfold of the chalk. On the south side of the London Basin the chalk forms a complete arch or anticline in the Hampshire part but east of this the top of the arch has been worn away exposing strata of the cretaceous, or chalk period older than the chalk itself, which occupy the Wealden area of Kent, Surrey and Sussex. The North and South Downs, the remaining fragments of the dome or arch of chalk, face each other across the Weald. South of the complete arch or dome of the Hampshire and Wiltshire chalk, of which Salisbury Plain is the broadest part, the chalk dips again in a trough or syncline of which the hollow is largely filled, as in the London Basin, by tertiary gravels. The most notable area is that of the New Forest and the Dorset Heaths, whose essential unity is apt to escape our notice on account of the county division which leads us to study

On the Norfolk Broads

The Reed Beds at Horning Ferry

them separately. Beyond these tertiary sands and gravels the chalk once more rises in a steep ridge which forms the backbone of the Isle of Purbeck and the Isle of Wight. A little consideration will soon make it clear that the line of the River Frome from Dorchester to Wareham, of Poole Harbour, Bournemouth Bay, the Solent and Spithead is the southern equivalent of the continuous valley of the Kennet and Lower Thames.

I. THE BROADS

The Lake District of the Cumbrian Mountains is not our only lake district, but that of the south-eastern quadrant of England is somewhat masked by the name of "Broadland". On the ten-miles-to-an-inch map the blue patches which show the Broads of Norfolk and the adjacent part of North-east Suffolk appear very small in comparison with the Cumbrian lakes; but the lakes of the Lowland are united by wide, slow streams, so that the "Broads", which include stream and lake, provide 200 miles of continuous waterway suitable for sailing, the largest area in England for inland cruising. Their scenery has not only the charm of space but also of colour, indeed on a sunny winter day there is no part of England which has a more beautiful scheme of colour than that of their deep blue waters fringed by wide stretches of yellow reed beds. This amphibious region is of particular interest to the naturalist as the breeding-place of rare birds; the higher ground which closely approaches the Broadland of the Bure displays noble churches and a quaint domestic architecture with the best of all kinds of thatching, that of the reeds. It is a fascinating district, but its beauty is vulnerable, for any incongruous erection is conspicuous in these open spaces; moreover, the charm of the place is largely dependent upon the quietude which harmonises with the visual character of the scene, and in default of restrictive regulations there is now no quietude in the summer holidays. In this connection it is well to remind ourselves of the fact that quietude and harmony of sound are among the amenities of the natural scene.

On several counts therefore concerted action on the part of local authorities is required for the preservation of amenities in the East Anglian Broadland, and it is desirable that such action should be undertaken with the advice and assistance of the archaeological and naturalist societies of the South-Eastern Union.

II. THE CORN LANDS OF EAST ANGLIA

The parish of Debenham at the source of the River Deben, in the centre of East Suffolk and on the clay soil, is situate in a typical part of the corn land of East Anglia. There are no villa residents, and, except for the fine churches, there is nothing striking to attract the tourist, for there is neither splendour of height nor of space, no grand feature of rock or charm of rapid stream. Yet for the resident, and I write as a former resident for here I was born and brought up, the social scenery is a perfect example of the Arcadian type for which England is famous. The small scale of the undulating landscape made the natural horizon more or less accord with the boundary of the parish, and this boundary roughly marked the limit of the daily trudge from the village to the fields. The village street was the factory of our local craftsmen. Local produce was sold in the village shops, and the windmills ground the corn we grew. Thus everything in the landscape was related to our life. We knew everyone by sight, and the part which he played in the economy of our existence. To the resident no kind of landscape has more human interest than that which displays the life of ordered agriculture, and nowhere does agriculture contribute so much to the scenic variety of the seasons as in the corn-growing districts. The ploughing and sowing, the growth and the reaping, mark the four seasons of the year; and the rotation of the crops imparts a yet longer rhythm to the colour pattern of the surrounding fields. Much of England retains this quiet charm which only the resident can fully feel, but it is a pathetic commentary on what is happening elsewhere that such districts are characterised by the significant word "unspoilt".

III. LONDON AND THE SUBURBS

The antithesis of the Arcadian scenery of the country parish is that of the national capital, situated in South-eastern England where the tidal Thames is first bridged. The capital proper is the "City" and Westminster. Here the features which illustrate the activities at the general headquarters of the nation are the great public buildings, from the Tower of London to the Houses of Parliament, from St Paul's to the Abbey, from the Guildhall to Buckingham Palace. There are also processional avenues of state pageantry and broad oases of parks, the whole constituting a fine material framework for a social life unsurpassed in wealth of intellectual interest and recreative variety. Yet throughout the lifetime of the eldest among us there has been a steady outflow from the vast built-up area of London town which encloses in a broad ring of continuous houses the central area of the Royal Court. This outflow has been in search of two things, fresh air and a garden. These advantages are enjoyed in the suburbs which spread radially on all sides of London, but only at a heavy price of social impoverishment. The very scenery of the suburbs is lacking in interest because the surroundings illustrate so little of the life of the residents. Further spreading of London's population would both increase this social impoverishment and impair the amenities of the surrounding rural counties. Fortunately there is a remedy in the replanning of the Metropolitan boroughs which surround the City and Westminster. Here are miles of streets of low houses of no architectural merit, and lacking modern conveniences. The same population as that now accommodated in this desert of bricks and mortar could be rehoused in steel-framed buildings on one-third of the space occupied by the present houses and the large area so gained could be laid out in boulevards, parks and gardens. Thus would London town become a garden city, and no city other than a garden city is really fit for human habitation.

This replanning would moreover much diminish the amount of smoke and therefore give freer access to the most rapidly vibrating of the solar rays. Thus through the agency of sunshine and vegetation the Londoner, reinvigorated by the vitamins of country air, would enjoy with a new zest that fullness of life which a great capital provides.

Access to the Metropolis has always been one of the advantages of residence in the Home Counties, but with the advent of motor traffic has come an attendant disadvantage resulting from the circumstance that London is by far the greatest radiant of arterial roads. There is no measure of scenic amenity more important than the prevention of building alongside these roads. I shall not, however, elaborate the point, for the matter is generally understood. What is needed now is the firm resolution to give full effect to such permissive powers as the forthcoming Ribbon Development Act shall put into the hands of Local Authorities.

IV. THE WEALDEN AREA

Between London and the coast to the south of it is a country of great natural charm which has unequalled advantages for those who require frequent but not daily access to London. The neighbouring part of the south coast is also a resort of Londoners on account both of proximity and advantages of climate.

Between the census of 1921 and 1931 there has been a great increase of population in the Sussex and Surrey portion of the Wealden area. The preservation of the picturesque old farm-houses and cottages is assured by the great demand for their occupation on the part of people of means who restore and keep them in excellent repair. It remains to ensure that new building shall be harmonious with this historic architecture; and this is now practicable, for our architects have devised a style of small house in brick and tile of sober tint and not too rigid in outline, taking its place quietly in the Wealden landscape. It seems to me that it is the old rustic architecture of the Wealden country which has provided the chief inspiration for this new type of small country house, which is, however, no mere imitation. Yet side by side with these excellent houses there are others, erected without professional guidance, which are both hideous in themselves and produce a horrible disharmony of grouping. I submit that we have reached a stage in housing when all buildings should be erected to the design of registered members of the architectural profession. We shall then be grateful to the building contractor who carries out in a workmanlike way the design which he is not himself in any way fitted or qualified to produce.

In the western part of the Wealden area the problem of by-pass roads is pressing, and on the adjacent coast the even more difficult problem of checking the erection of ugly shacks and vulgar bungalows. In these districts are a number of societies affiliated to the South-Eastern Union, and I venture to suggest that their most practical means of action would be the selection of members possessing the administrative mind to stand for election on the Urban and Rural Councils now in possession of powers under the Town and Country Planning Act which they are at liberty to use or not to use. The will to exercise these powers needs strengthening by an increase of the scientific and artistic element on the Councils.

V. THE LONDON BASIN AND THE DISTRICT OF THE FORMER WINDSOR FOREST

Central in the London Basin is the area which was formerly covered by Windsor Forest. Both in size and in the character of soil this area bears a considerable resemblance to the New Forest in the geologically similar Hampshire Basin. Bagshot Park, the former Royal hunting lodge and still a residence of royalty, occupied the central position in the forest as the boundaries ran in Stuart times. A circle of 10 miles radius round Bagshot included most of the forest and not much that was outside. The boundary was the Thames from the junction of the Loddon, near Wargrave, to Weybridge, then along the River Wey to Guildford, the Hog's Back to the Blackwater, the Blackwater to its junction with the Loddon and then by the Loddon to its junction with the Thames, which completes the circuit. The portion of this area which is nearest to Windsor Castle is preserved as the park. The rest of the fertile land is farmed, but a large proportion of the area is barren sand and gravel. A considerable part of this is now "Crown Land", a very misleading description of land which has been taken over by the nation from the personal possession of a sovereign. This unfortunate misnomer is certainly responsible in no small measure for the supine attitude of the public towards the action of the Commissioners of Crown Lands in converting parts of this delightful wilderness into golf links and in offering other parts for villa development. Any argument that such a policy is justified on economic grounds is entirely out of date in view of the need for wild open spaces in the environs of our rapidly growing metropolis and the great expense which would attend the purchase of land for the purpose. How urgent this need will soon become is evident from the startling fact that the population of London, and the neighbouring urban areas which are economically part of London, has increased by more than a million since the Great War, and that this rapid increase still continues.

It happens that the Royal Fine Art Commission has recently been given an advisory position in relation to the Commissioners of Crown Lands in matters of amenity, and I venture to urge that the Fine Art Commission should promptly take this matter up and advise that the remaining space of heath and pine in the area of the old Windsor Forest, now under the control of the Commissioners of Crown Lands, should be permanently maintained in the wild state.

VI. THE NEW FOREST

The scenery of the New Forest is of great importance to the nation as the principal remnant of the characteristic English woodland of broad-leaved, deciduous trees. The rights of the commoners have done much to preserve the charm of this legacy of Old England by maintaining part as open forest, but the beauty of the enclosed forest has been greatly impaired by the planting of pine trees. The Forestry Commission of Great Britain, which now has control, seeks a reasonable compromise between economics and amenity in the areas under its administration,

Arne Chapel

By Poole Harbour

but I urge most strongly that in the New Forest, which is in effect our first national park, the claims of amenity are paramount and that no more pine trees should be planted. It is true that in its proper place and setting a venerable pine is beautiful, but scenic beauty is mainly dependent upon harmonious grouping, and no grouping is more unsuitable than a formal pine plantation amidst woodlands of British oaks.

VII. THE HEATHLANDS OF THE ISLE OF PURBECK

Christchurch, Bournemouth and Poole lie between the New Forest and the heathlands of the Isle of Purbeck, and the preservation of both these districts in their natural state is of great importance to the residents and visitors of the great conurbation, or group of towns. Between the southern shore of Poole Harbour and the road connecting Studland with Wareham lie Studland Heath, Godlingstone Heath, Newton Heath, Wytch Heath, Norden Heath, Middlebere Heath, Slepe Heath and Arne Heath. Between the heaths and the sandy foreland of Studland Bay lies the lake called Little Sea, a breeding place for waterfowl. All this wild heathland is in striking contrast with the populous northern shore. The shallow, intricate and unfrequented channels which lead to the inlets of the southern shore of Poole Harbour are also in strong contrast with the frequented channel which leads to the Port of Poole. At the time when the Isle of Purbeck was settled by the Saxons, and throughout the Middle Ages the conditions were the reverse of these, and Wareham was the principal port until the southern channels became silted by the sediments brought by the Puddle, Frome and Corfe rivers. There were also other landing places, as at Arne Bay where the beautiful old chapel, mounted upon a detached knoll of plateau gravel, is now as a voice in the wilderness. Encircling the heaths at Studland, Corfe and Wareham are noble monuments of the distant past. The importance of preserving in its wildness the scenery which lies between these towns and Poole Harbour is evident to those who have taken note of the recent phenomenal growth of population of the continuous built-up area of the boroughs of Bournemouth, Poole and Christchurch. Of the seventy-four conurbations, or urban groups, which are enumerated in Professor C. B. Fawcett's valuable paper on "The Distribution of the Urban Population in Great Britain,"[1] that of Bournemouth, Poole and Christchurch had the greatest proportional increase in the last decennial period, no less than 25 per cent., the total reached in 1931 being 183,000. Before the rise of Bournemouth as a watering place the heathland on the further side of Poole Harbour was of little aesthetic value, being an inconsiderable part of a wider wilderness. But such a stretch of utterly elemental scenery of land and water immediately adjacent to an urban population already verging on 200,000 is very precious, for it provides a place of contemplation where man may recover that sense of intimate communion with Nature which is a pure and undefiled source of religious inspiration. So little teaching of this kind comes from the pulpit that many a devout soul is never brought to realise

[1] *Geographical Journal*, Feb. 1932.

the inspirational value of that sense of peace amidst the surroundings of untouched Nature which comes to a mind harassed by the turmoil of modern life. But although the average busy man may not have realised the revelational value of wild scenery, he does crave for its quietude, and thus the preservation of such a district as I have described is of economic value to a neighbouring urban centre. This open heathland is very vulnerable to disfigurement, for any shack or bungalow stands out conspicuously. I contend therefore that the whole area should be preserved in its pristine state by the co-operation of the municipality of Poole and the County Council of Dorset. In this scheme, the Bournemouth Corporation might well take a friendly and helpful interest.

We have reached a stage in the suburbanisation of the south-eastern quadrant of England which calls for expenditure in the compensation of landowners much greater than that to which our minds are yet accustomed. If, however, the rate-paying public concerned will master the statistics of the recent and present growth of population in these parts of England they will realise that the sums required are now spread out over such a large and growing community that they are far less formidable than they appear to those who only carry in their minds the figures of pre-war population.

VIII. THE CHALK DOWNS

The long chalk ridge which forms the backbone of the Isle of Purbeck bounds the panorama of Bournemouth Bay on its western side and thus its heights suggest an excursion which is in fact always well rewarded, for the downland pasture displays the chalk formation at its best. Chalk rock is sufficiently friable for the summits to be gently rounded, sufficiently compact in the mass to stand at a steep slope on the flanks, and being soluble in rain-water the troughs can be deepened without need of a sloping side channel. The result is an unbroken continuity of curve not equalled I believe by any other rock, so that the chalk downs have a form very similar to the surface of an ocean heaving with the undulation of a crossing swell. Time was when most of the chalk country was grassy down, but in the eighteenth century the broad and less steep parts began to be ploughed and there are now very large areas of arable land on the chalk where the delicate curves into which it has weathered are camouflaged by the fences and divisions of the fields. Generally speaking it is the steepest slopes of the chalk country which have escaped the plough. The turf-covered downs which provide the most beautiful upland scenery of South-eastern England are therefore situated on the narrow ridges, as those of Purbeck and the Isle of Wight; on the long course of the north-western escarpment which winds through Dorset, Wiltshire and Berkshire, with some outliers in the Chilterns as Ivinghoe Beacon, and lastly in the South Downs of Sussex. The scenic importance of these long lines of chalk pastureland is twofold, that of natural beauty and prehistoric interest. The charm of the suave lines and the distant prospect over broad vales, as the Vale of White Horse and the Vale of Aylesbury, is enhanced by the freshness of the downland air deliciously scented by the wild thyme, and the pastoral associations of the shepherd and his flock harmonise with the natural background. It is, moreover, mainly on these steep heights of the chalk downland and overlooking the broad vales that the great fortified camps of prehistoric times are situated. Their positions are frequently some of the finest view-points in South-eastern England, suggesting that the advantage of the height had at least as much to do with a commanding outlook as with difficulty of ascent. It is evident that what little of the chalk still remains in natural pasture should, on account of its natural beauty, be protected and maintained in its present condition under the powers provided by the Town and Country Planning Act. The importance of this step is, moreover, enhanced by the fact that only thus can we ensure that the wonderful prehistoric camps, belonging to an age which is now being elucidated by archaeological research, shall be preserved in a setting suitable to the conditions of their original construction. It is not too late to save them.

The megalithic cathedrals of prehistoric England, Avebury and Stonehenge, are on the chalk plateau which has been ploughed up. The surroundings are thus modern in character, but under the powers of the Town and Country Planning Act it is, at all events, possible to regulate any further development of their neighbourhood in such a manner that the surroundings shall remain purely agricultural and be kept free from mechanical industries and exploitation for the purpose of sight-seeing.[1]

[1] See "Avebury, a great prehistoric site," by the Rt Hon. W. Ormsby Gore, *The Times*, May 31st, 1935.

The natural beauties and archaeological treasures of the chalk uplands are spread over many counties, but come within the area of the South-Eastern Union of Scientific Societies and that of the neighbouring South-Western Naturalists' Union. It seems desirable therefore that these two bodies should approach the local authorities with the object of securing a consistent plan for the preservation of scenic amenity throughout the wide ramifications of these highlands of the south-eastern quadrant of England.

IX. THE CLIFF LANDS

From Kent to Dorset a succession of long ranges of chalk downs terminates abruptly in perpendicular cliffs of important height. Views such as that from Beachy Head and from the downs between Freshwater and the Needles, are extraordinarily impressive, largely owing to the circumstance that the sheer descent so lures the eye that although the sea horizon is the focus of attention it is not in the centre, but near the top, of the picture, thus imparting an appearance of immense expanse to the sea; an example of the unconscious artistry of the eye.

It is some time since the authorities of Eastbourne prudently acquired possession of the cliff lands of Beachy Head. It is very desirable that, even where the rateable value of a district is not sufficient to enable the authority to acquire the back lands, a strip of about 100 yards from the cliff edge should be protected not only against building but also against enclosure by a garden fence. Such a precautionary measure is necessary to ensure the full benefit of the sea view from the cliffs, for this wide and noble prospect cannot be properly enjoyed if the spectator is cribbed and confined on the landward side.

I trust that the members of both the South-Eastern Union and the South-Western Naturalists' Union will use their influence to secure an open, unenclosed pedestrian broadway 100 yards in width along all the lofty cliffs of Southern England. Let it never be forgotten that the sea view from the cliffs is the special scenic heritage of our island people.

CHAPTER V

THE CLIFF LANDS OF ENGLAND AND THE PRESERVATION OF THEIR AMENITIES

GEOGRAPHICAL SECTION, BRITISH ASSOCIATION, NORWICH MEETING, 6 SEPTEMBER 1935

(*From* The Geographical Journal, *vol.* LXXXVI, *no.* 6, *December* 1935)

The view of the sea from the beach is an unfailing harmony of tone and colour, for the water responds instantaneously to the changes in the sky. Ascend the cliff and to these charms is added the inspiring impression of immensity. In this effect two causes co-operate. With the increased elevation of the eye more of the watery plain comes into view, a narrow band indeed but bringing that softness of tone which suggests great distance, an effect which is often enhanced by the close clustering of the yet more distant clouds which come into sight above the delicate line of the horizon.

The impression of immensity in the prospect from the cliff does not, however, depend wholly upon height but in great measure upon the steeply descending foreground. This is due to the fact that although we detect at once a lateral slant in the view, the human eye is not sensitive to a departure from horizontality in the direction of its outlook. Even when the horizon is upon the same gravitational level as the eye, or only differs from that level by the small angle due to the curvature of the earth, the position of the horizon in the field of vision depends mainly upon the slope of the foreground. When standing upon flat fields of the Fen country or on Romney Marsh the ground-line 10 degrees below the horizon is within 30 feet of the observer and nothing much nearer than this can be focussed at the same time as the distant view. There is, moreover, a sort of compulsion upon the subconscious mind to reject the idea that the paltry distances within a few steps of our standpoint are a large part of a full-scale landscape. Ten degrees is not nearly the half of our natural field of scenic vision and so in the flat land of fen or marsh the plane of sight habitually inclines upwards, we take into our view much more of the sky than of the land without any warning sensation of the upward slant of the eye, and so we all feel that the sky of the Fen country is splendidly spacious. This is due to the fact that we look so as to see more, not because there is more to see.

Human stature is puny in comparison with the distance of the horizon given by the great radius of the sphere upon which we live, and so when standing on the fen or beside the sea the terrestrial view is inconveniently foreshortened. In Brobdingnag, as we know from Gulliver's narrative, the stature of a man is ten times as great as in England, but the distance of his horizon is only increased threefold and the proportions of the terrestrial part of the field of view are better than they are for us. But we have only to take our station on the verge of a lofty cliff and we obtain at once the scenic outlook of a more than Brobdingnagian stature. The watery plain 30 degrees below the gravitational level is now sufficiently far off to be focused with the horizon, and as viewed from a giddy height the diminished pattern of the waves below appears to be at a considerable distance. From such an elevated level of the eye a super-Brobdingnagian might view the horizon in or near the middle of the picture, but to a mere man suddenly raised to a stature of 200 feet the view below the horizon is the more important part of the picture and the direction of his outlook inclines downwards spontaneously and without any muscular sensation. The sky view is thus diminished, the horizon comes near the top of the field of vision and the watery plain embracing nearly all the field appears vast indeed.

There is no mathematical formula to enable us to evaluate the two factors of height and steepness which contribute to the impression of space in the view from a steep and lofty coast and I have therefore made observations in different places to obtain a standard by which to judge what parts of the English coast are to be included in those cliff lands which provide an impressive prospect of the sea. From a height of about 100 feet on the verge of the steep cliff west of Branksome Chine on the way to Poole Haven the apparent expanse of sea is notably enlarged by the raising of the horizon in the field of vision and there is a sense of formidable depth below. On the east of Branksome Chine, that is to say on the Bournemouth side, there is a short reach of cliff of only 50 feet. From this the outlook on the bay and its coast is charming and beautiful, but the prospect of the sea itself is far less impressive than from the 100-foot cliff. From this and confirmatory observations I adopt 100 feet as the least

Beachy Head

The raised horizon of the sea

height for the cliff lands which are to be included in our scenic survey.

Next as to the factor of steepness. A perpendicular cliff when viewed in profile is much more impressive than a sloping cliff, but when standing on the summit of an almost perpendicular cliff, as at the end of Berry Head in South Devon, it is not practicable to view the face, and, no recessional plane being seen, some of the visual effect of the gravitational descent is lost. Thus the summit of a boldly sloping cliff whose declivity can be seen as the foreground of the view provides a more impressive outlook station.

In seeking to determine the order of steepness which is thought of as a bold slope when viewed from the summit I shall cite the case of Salcombe Hill on the east of Sidmouth in South Devon. The sea view from the summit (535 feet) is very impressive, and from the verge of the cliff the descent appears to be of formidable steepness. The general slope from the 500-foot contour to high-water mark is about 40 degrees as measured on the 6-inch Ordnance Survey map.

I have found in many localities that a cliff appears to be of formidable steepness when the base is two spans of the hand (held at arms length) below the sea horizon, that is to say, in rough measurement 36 degrees.

When we stand on a cliff the fact that the sloping plane is visually ascending all the way as the eye travels from the summit to the beach is difficult to realise, and I have found by indirect measurement that the visual ascent is enormously underestimated in the spontaneous impression of the scene.

The view-points on the English coast from which the sea horizon stands high up in the picture and the scene has the impressive character associated with a cliff prospect are not limited to the immediate vicinity of the edge of broken rock in those localities where the hill rises steeply from the cliff summit. Thus near the Signal Station on Beachy Head (525 feet) where the grassy down slopes smoothly at an angle of about 25 degrees to the cliff edge, the sea horizon is almost at the top of the field of view as recorded in a sketch which I made from this position.

The case of the sea view from Heytor Down in Southern Dartmoor, 1100 feet in height, is very different. This is 11 miles from the coast at Teignmouth, so that the general slope is little more than 1 degree. The sea is visible above a gently descending landscape. In two sketches which I made from different standpoints the horizon of the sea is slightly below the middle of the picture. Although the sea horizon was 41 statute miles away and almost 30 miles of water were included in the view the expanse of sea was not impressive, much less so than from the railway train when passing along the level shore between Dawlish and Teignmouth where the horizon is about 5 miles distant.

As we go farther west along the coast we reach the hardest rocks, primary metamorphic and igneous. On the long line of the cliffs in South-west Devon and in Cornwall there are places where the waves and weather have only succeeded in cutting a cliff of less than 100 feet in height yet the steepness of the hillside ensures an impressive character which is not displayed where cliffs of less than 100 feet have a plateau summit. Hope's Nose near Torquay is a familiar example. From this and other cases I venture on the generalisation that the coastal outlook has the impressive character of cliff scenery when the land reaches the 200-foot contour at a distance of 660 feet (1 furlong) from high-water mark, even when the face of broken rock is less than 100 feet.

Measured on the 1-mile-to-1-inch Ordnance Survey maps[1] the length of coast with a cliff not less than 100 feet, or alternatively attaining a height of 200 feet within 1 furlong (660 feet) from the shore is 530 miles out of a total length of 1800 miles. Thirty miles are crumbling gravels of the tertiary formations. Some of these, as along the Christchurch and Bournemouth Bays, have real beauty, partly on account of the sweeping curve of the coast, but because of their rapid wastage I omit them from my computation of the cliff frontage where a strip of top land should be purchased. This leaves 500 miles as the length of cliff lands which have to be considered in reference to the project of public ownership. Of this 500 miles the chalk cliffs in short stretches between South Yorkshire and East Dorset account for 51 miles. The jurassic-triassic rocks account for 126 miles, forming two belts, each continuous, one in North Yorkshire, the other in West Dorset and South-east Devon. The lengths of the two belts are approximately equal. Finally the hardest rocks which

[1] This is a first approximation which gives us the order of magnitude with which we have to deal. Measurement on the 6-inch-to-1-mile map would be preferable, but, as my main purpose is to estimate the area of cliff land required for reservation, minute indentations are of comparatively small importance.

form most of the northern and southern coasts of the south-western or Devonian peninsula (meeting at Land's End) account for no less than 323 of the whole 500 miles of English cliffs.

Chalk rock, although not a hard substance, is compact in structure and strong in the mass. It therefore stands in nearly vertical cliffs which are often capped with steeply sloping turf as in the cases already mentioned.

The cliffs comprised in the geological categories of jurassic, liassic, and triassic are composed of oolite limestone, lias, red marl, and new red sandstone. This belt of rocks traverses England diagonally and its truncated ends appear on the east coast from Saltburn to Filey Brig, and on the south coast from Swanage to Teignmouth.

Boulby Head, north of Whitby (666 feet), is reckoned to be the highest cliff on the English coast. On the Dorset-East Devon front of the jurassic-triassic shore there are also many lofty cliffs as in the long line of heights from Seaton to Sidmouth.

The land on the top of the lofty cliffs of West Dorset and South-east Devon is for the most part of a plateau character and this is typical of high coasts where the strata are soft, the rock breaking back not merely where undermined by the waves and tides but where exposed to the salt sea winds which prevent the growth of sufficient vegetation to bind the surface.

After passing Teignmouth the coast of the English Channel from Torquay to Land's End continues to be mainly cliffs, but here the physical character of the rock is very different. The sedimentary rocks of the carboni-

The Granite Cliffs of Cornwall
From Gurnard's Head to Pendeen Watch

ferous and Devonian periods, which occupy the greater part of the coast, are much harder than the more recent sedimentary rocks from Yorkshire to South-east Devon, and even harder than these ancient sedimentary rocks are the metamorphic and igneous rocks which stand out at the promontories of the Start, the Lizard, and Land's End. The profile of this coast is characterised by a cliff face backed by a steep slope rising to the summit of hills of a few hundred feet only in height. The impressive character of the actual cliff is not dependent solely upon the attributes of height and precipitance, being everywhere enhanced by the evident strength of its substance. In structural character the rocks are of two principal classes, the slates in laminae and the granite in massive blocks. The slates of the south coast of Devon and Cornwall reappear on the north coast of the Devonian peninsula, but not elsewhere on the English shore. Their effect upon the marine outlook is largely a matter of tone and colour. In South-western Cornwall and on the north coast of Cornwall the sea is of the oceanic blue. Near the Lizard the slates, which are often almost black and of neutral tint, serve on sunny days to carry depth of tone farther than that reached by the ocean blue. Between Tintagel and Boscastle on the north coast there is an even finer effect. Not only is the water itself, as I judge, of a somewhat deeper blue, but the slate is of a dark blue tint, and, lighted from pale blue sky and deep blue sea, carries the chord of colour down to a depth of tint rarely met with in the English scene. Thus on the stretch of Cornish coast which legend associates with King Arthur there is a scheme of colour which fully equals that which can be found beneath Italian skies or by the Mediterranean shore.

Between the slate cliffs of the south and north coasts of the south-western or Devonian peninsula juts out the promontory of Land's End, a dome or boss of granite, one of a succession of igneous upthrusts of which other members are Dartmoor, the Bodmin Moors and away to the west the Scilly Isles. The Land's End peninsula, however, is the only mass of granite upon the coast of England, and so the granite cliffs which extend for many miles between Penzance and St Ives are a unique feature in English scenery. The "Land's End" in the popular sense, the terminal point of the extensive peninsula, is visited by people from all parts of the country, and the scene does not disappoint the expectations raised by the romantic suggestion of the name. The granite, here bedded horizontally, rises in perpendicular cliffs and is so jointed in the horizontal and vertical planes that it has the form of cubic blocks. Thus England's farthest outpost on the Atlantic resembles a defensive rampart piled up by giants against the onset of the sea.

In order to frame a satisfactory scheme for the scenic preservation of these cliff lands the problem must be studied in accordance with the method of geographical science, that is to say, both from the general and the regional standpoint. Regionally considered, it is evident that the support for the project suffers from the great distances which separate the residents of our coast lands. North Yorkshire and South-east Devon have similar cliffs, but the communities concerned live far apart and are entirely unconnected in their local politics. When, however, we survey the coastline from the general or national standpoint we cannot fail to be impressed by the importance of the fact that no town in England is more than 75 miles from the sea, that is to say not more than three hours by car, even for conscientious people. The sea-coast, so readily accessible from all parts of the country, provides the most beneficial climatic change obtainable in England, and is the principal holiday resort. The preservation of a belt of unspoilt land along the cliffs is therefore of great importance to the general public as well as to seaside residents, and should be made the joint concern of His Majesty's Government and of Local Authorities.

I shall assume that Local Authorities will fulfil the obvious duty of securing a public footpath along all the cliffs which are not already enclosed to the edge in a private garden. But this is not sufficient to secure the amenities of cliff scenery, for the splendid spaciousness of the ocean view cannot be fully enjoyed unless the landward side be free and open. With regard to the further steps which must be taken in order to secure the amenity of the cliffs I may refer to a planning scheme of one of the Local Authorities which requires the building line to be set back 100 yards from the cliff. But if the garden is to be in front of the house, the cliff path will be hemmed in by the garden fence. I submit that the essential provision is that the frontage of the building plots should be set well back from the cliff, and that the belt of land thus kept open should be acquired as a public pleasance.

I come now to the calculation of the acreage of such a pleasance extending for the whole 500 miles of the English cliffs. In order to obtain the simple figures which are necessary for memorisation and therefore for clear thinking we must employ the customary unit of English land measurement which is very characteristically the length of the cricket pitch. The distance between the wickets, 22 yards, is determined by the length of that antique instrument the chain. I take 5 chains (110 yards), as a suitable breadth for the proposed pleasance of the cliffs. Ten square chains are 1 acre, so that each acre of this pleasance has a cliff frontage of 2 chains (44 yards). This is one-fortieth of a mile, so that the proposed pleasance has an area of 40 acres per mile and the whole national reservation extending along the 500 miles of English cliff an area of 20,000 acres.

Let us compare this with the area of the districts most favoured for acquisition as national parks. The New Forest has an area of more than 64,000 acres, and the proposed national parks of the Lake District and Snowdonia are on the same scale, that is to say more than three times the size of the proposed reservation of the cliffs.

Under the provisions of the Town and Country Planning Act, 1932, Local Authorities have the right to purchase land for public use. Their power to do so is limited by rateable value. Where the cliffs come within the boundaries of the larger seaside towns the local resources are ample for the purpose, and the convenience of undivided control would probably lead the Local Authority to prefer acquisition without contribution from the Exchequer. Matters are very different on the long stretches of coast comprised in rural districts, and it is precisely these less populous parts of the cliff coast, especially in the rural county of Cornwall, which are of most importance for national recreation. It would seem therefore that for the acquisition of cliff lands for public use on the less populous parts of the coast it would be entirely just and right that there should be a contribution from the National Exchequer.

Let us consider on the other hand the consequences of a policy of *laissez-faire*.

Much of the 500 miles of cliff land is extremely attractive for house building on account of the lovely view, and this applies now more than formerly because of the great increase of holiday residences, week-end houses, and camping huts. When the lovers of scenery revisit some favourite cliff walk and see that building has begun, they throw up their hands in horror and rail against the mercenary spirit of people who wantonly destroy England's beauty. But this attitude of mind shows a want of imagination. Let the visitor put himself in the position of the small landowner deriving a meagre return from the produce or the rent of cliff land, which is mostly of small agricultural value. When he is offered a building price for plots of land he sees his way at once to easier circumstances of life, and let us not forget that when financial circumstances become easier the world at once assumes a cheerful aspect and the eye becomes more tolerant of disharmonies in scenery. Moreover, in rural districts it is not only the landowner who welcomes the extension of villadom to cliff lands, for the arrival of even a dozen new families in the neighbourhood makes a difference to the village shopkeeper.

Where, on the other hand, the cliff fields are now part of a large property the inducement to sell for building purposes will be very great at the time of the next succession as a means of meeting the death duties which are levied upon large properties on a destructive scale.

Taking 1 chain (22 yards), as the building frontage of the plot required for each modest seaside villa, there would be 80 houses per mile, or 40,000 for the whole 500 miles of the cliff coast. The report of the Ministry of Health shows that the number of houses built in England during the six months ending 31 March 1935 was no less than 160,000, so that it would be rash to cherish the hope that a long time would elapse before the wild cliff lands of England could be transformed into mere villadom.

Pending a valuation survey, the following calculation will serve to give an idea of the scale of expenditure required for the acquisition of the cliff lands for public use, and the relation of that cost to the expenditure upon other public undertakings of a cognate character. The agricultural value of cliff land is generally small, the price of a building site for which water, drainage, and other services have been provided is high, but the cliff lands as a whole have at present no such services. Account must, however, also be taken of the value of sites for camping and for restaurants. Basing my estimate upon some actual cases I take £100 per acre as a first approximation for the value of cliff lands outside the boundaries of prosperous watering places. Twenty thousand acres at £100 per acre is two million pounds. The comparison which I suggest is with the cost of slum clearance, for slum clearance and the preservation of wild scenery ought not to be put into entirely separate categories. On the contrary, the project for slum clearance and that for national parks ought to be envisaged together as complementary parts of one great movement for saving England from what is mildly termed undue urbanisation, a condition that is to say in which towns are not fit to live in and the countryside not fit to look at.

It will suffice for our purpose to quote the proposed expenditure by the London County Council on slum clearance and rehousing which amounts to a capital sum of no less than thirty-five million pounds.

In 1929 the Government appointed a committee to consider the advisability of establishing national parks under a national authority. Upon the evidence of the expert witnesses, of whom I was one, the committee recommended that a national authority should be constituted for the purpose, and national expenditure authorised. The Government, however, have not given effect to the recommendation of their committee.

The Town and Country Planning Act, 1932, has modified the problem of national parks by giving Local Authorities some of the powers for acquiring reservations of scenery for which it had previously been considered that a national authority would have to be constituted. Nevertheless there is still the need for a national authority furnished with funds and charged not only with duties of supervision in matters of scenic preservation but with the obligation of initiative. There can in fact be no proper planning for the preservation of English landscape whilst the only competent authorities are those whose outlook is necessarily restricted to a single corner of the country, and this applies particularly to the case of the cliff lands.

CHAPTER VI

THE PRESERVATION OF NATURE IN ENGLAND

GEOGRAPHICAL SECTION, BRITISH ASSOCIATION
BLACKPOOL MEETING, 14 SEPTEMBER 1936

I. INTRODUCTION

The scenery of civilisation is composed of two parts, the works of Man and the natural background, and the scenic heritage of every nation depends for its value upon the maintenance of due proportion between these parts.

The spread of building and increase of mechanical traffic in England have made the public realise the need of planning for the preservation and even for the restoration of the natural background, and the passing of the Town and Country Planning Act of 1932, and the Restriction of Ribbon Development Act of 1935, have placed in the hands of Urban and Rural Authorities powers which are of a novel character. Having regard to the scope of the work hitherto undertaken by these bodies, it is not surprising that few of their members are experts in natural science or have made a special study of the principles which govern the aesthetics of scenery. On the other hand, we have in England a great number of learned societies, whose members possess the requisite knowledge but no administrative authority. It is desirable therefore that Local Authorities and all persons interested in local planning schemes should get into touch with these societies through the good offices of the British Association for the Advancement of Science and the Council for the Preservation of Rural England, two bodies with which most of the societies in question are in some way affiliated or connected.

II. THE RESTORATION OF NATURE IN THE MODERN TOWN

A scene from which Nature has been expelled is no fit dwelling-place for Man, yet such is the modern industrial town where so large a proportion of our population lives. The greater part of London, outside the West End, comes under this category. Considered merely as a picture an unrelieved expanse of stone and brick is insufferably harsh and forbidding; moreover, in smoke-laden atmosphere the people are deprived of that fresh and fragrant air which is the rightful heritage of those who are native toEngland's green and pleasant land. Flight to the suburbs is lamentably destructive of the civic sense. Therefore let every practicable means of smoke abatement be promoted by the authorities; let every householder brighten the view with flowers in window boxes whose colours are so telling against the background of masonry; and above all let the rebuilding schemes be for the accommodation of no more than the present population in higher houses occupying less land and thus leaving space for town gardens and boulevard avenues.

The Roman Wall

The North Gate, Housesteads

III. A GREEN BELT, AND THE PREVENTION OF RIBBON DEVELOPMENT

The maps prepared by Dr P. W. Bryan of the University College, Leicester, on which every building is shown and all other details omitted present an astonishing picture of the large proportion of England which is already embraced by the octopus arms of radial suburbs and by that unsightly intrusion on the countryside for which Professor H. J. Fleure coined the term "ribbon development". The preservation of a green belt round the cities, the substitution of satellite towns for radial suburbs, and the general prevention of ribbon development both beside roads and along the coast are primary tasks in the preservation of a natural background for the architectural features of our scenery.

IV. THE PRESERVATION OF A CONTEMPORARY BACKGROUND FOR THE MONUMENTS OF ANTIQUITY

We have now not only a Society for the Protection of Ancient Buildings but in the Office of Works a Government Department capable of enthusiasm for academic things. But for the preservation of historic scenery it is not only needful that the historic structures should be preserved but that they should be viewed in a setting concordant with the conditions under which they were erected. In our towns this is not possible, and rarely in our villages, so that mediaeval, renaissance and eighteenth-century architecture must as a rule be viewed together with modern buildings. It is otherwise, however, with the monuments of British and Roman times. Many of the most important examples of the former are situated upon the chalk uplands of South-eastern England, still scarcely inhabited and of which a considerable part is open pasture. Here are numerous earthen encampments in conspicuous sites crowning steep escarpments, which command glorious views. It is one of the firstfruits of the Town and Country Planning Act that arrangements are being made between landowners and Local Authorities for the preservation of pastoral conditions in several of these areas.

Even more precious than the encampments are the megalithic remains in the chalk country of which Stonehenge is a pre-eminent example. Through the open colonnades of this round temple of the sun the rising and setting of the giver of light and source of fertility were seen at all seasons of the year beyond the smooth horizon of Salisbury Plain, and only when the shrine of Nature worship is viewed in this appropriate setting can its majesty and beauty be fully realised and felt. Something has already been done to restore the surroundings to the pristine simplicity which was so shockingly impaired by the heavy hand of the War Office.

Remote from these prehistoric monuments of ancient Britain, amidst wilder scenery than that of the rolling downs, the long rampart of the Roman Wall crowns a rugged outcrop of volcanic rock where lofty, barren moors form the chine of Northumbria between the North Sea and the Solway Firth. The camp at Housesteads, with walls of well-squared stones and pavement scored by the ruts of chariot wheels, is the centre of the scene and

The Whin Sill Crag

At Housesteads by the Roman Wall

commands a northern view as stern and wild as in the days of the heathen Pictish tribes. This great monument of Hadrian's reign is not only our national heritage but a trust held by us on behalf of many nations, for it is a priceless relic of the defensive frontier of the city civilisation of the ancient world against the nomad tribes of the North, a frontier which stretched across Europe and Asia from the Atlantic to the Pacific Ocean. Housesteads and its immediate surroundings are already protected and preserved through private generosity and the careful provisions of the Office of Works, but more remains to be done, and the poverty of the land, both as regards fertility and game, makes the task comparatively easy. Here the reservation from all building should be on the scale of a national park so that this monument of Roman times, unrivalled elsewhere in Britain and significant alike in the history of England, Wales and Scotland, should always be viewed as it is to-day in a setting which conforms with the purpose for which it was erected and the conditions which obtained at the time of its origin.

V. THE COUNTRY OF THE BROADS

When we leave the lofty downs of the chalk country and find ourselves surrounded by the insignificant eminences of East Anglia the sense of grandeur departs, yet that noble attribute unexpectedly returns when these slight inequalities are smoothed out and we reach the open expanse of the Broads and the neighbouring flood-plains of meandering rivers, round which in peaceful silence the slow current winds. The simplicity of line and openness of view make these scenes particularly vulnerable to the jarring effect of shacks and bungalows. Moreover, the circumstance that smooth, inland waters lend themselves to holiday pursuits has brought the blare of gramophones to the once silent Broads. The emotional appeal of Nature has to do with the conjoint impression of the several senses, not with that of sight alone, and the time has come when by-laws are needed for maintaining peaceful quietude in places whose natural beauty can only exercise its healing effect when freed from the vulgarity of noise.

VI. THE PRESERVATION OF WILD FLOWERS AND THE ESTABLISHMENT OF RESERVES FOR THE PROTECTION OF RARE SPECIES OF PLANTS AND ANIMALS

It is common knowledge that the introduction of the motor-car has increased threefold or fourfold the radius of daytime excursions from the towns, but in reckoning the result in regard to the gathering or uprooting of wild flowers we are apt to forget that the area so visited is proportional to the square of these numbers, so that the country liable to this devastation is about twelve times as great as formerly. By-laws may be enacted but can so seldom be put in force that the co-operation of educative bodies, as Flora's League, should be sought by Local Authorities. Broadly speaking, the aim should be to raise the instinctive love of flowers as decorative objects to the level of a sympathetic appreciation of the flowering plant as a living thing which should not be wantonly destroyed.

Those species of wild plants already rare are not so

much in danger from childish ignorance as from the greed of collectors, the criminal class of Nature lovers. The small areas where the rare species grow and from which their seeds can spread have to be protected as Nature Reserves.

Rare species of insects have to be protected in like manner, as in Wicken Fen, between Cambridge and Ely, where the swallowtail and other butterflies breed and send out whole regiments which, spreading over the country, bring new forms and colours displayed with the added charm of wayward flight.

The reserves or sanctuaries for the breeding places of birds have to be on a larger scale and in isolated situations, as Blakeney Point and Scolt Head in Norfolk and the Farne Islands off the coast of Northumberland. These and others are in the safe guardianship of the Society whose full title is the National Trust for Places of Historic Interest or National Beauty.

VII. ON THE PUBLIC OWNERSHIP OF CLIFF LANDS

When we compare England with the principal countries of continental Europe in respect of wild scenery, it is at once apparent that we have relatively little of mountain or forest but excel in the extent of our coast which is no less than 1800 miles in length. Here the prospect is for all time that of elemental Nature. No part of England is more than 75 miles from the line of breakers where the waves of the ocean reach the culmination of their beauty in the rush of the curling crest. And to the wanderer by the waves our tidal shore offers more attractions than the narrow Mediterranean beach. The strand which lies between the wrack of seaweed and the line where the waves lap at low water has many allurements. Here the rock pool with seaweeds seen through clear salt water excels in lustrous colour all the gardens of the land.

Moreover, England's shore has 500 miles of cliffs, and the view from their summit stimulates the imagination to realise in full measure the immensity of the ocean.

The coast is the chief holiday resort of England alike for townsmen and countrymen, for here is not only change of scene but beneficial change of air. The tidal strand beyond the frontage of the towns is fairly safe, but there is immediate danger that the cliff lands with their unrivalled view may be lost to the public. In times past, which in reference to scenic preservation means before the advent of the motor, a few houses only here and there had private grounds reaching to the verge of the cliff, thereby interrupting the path of the pedestrian, and elsewhere the cliff path was usually open to landward, not hemmed in by a barbed-wire fence. Now all this is changed. Holiday bungalows are built, shacks put up, camping grounds reserved, and on many cliff lands the number of visitors is so great that either their wandering on the farm land must be restricted or the land acquired for a public open space.

The length of the cliffs, 500 miles, is great but the breadth of cliff land which needs reservation as an open space in order to preserve the character of the view is comparatively small. A strip of 5 chains (110 yards), wide I estimate to be sufficient except where the land rises somewhat steeply from the pitch or broken edge of the cliff.

In framing a policy for placing cliff lands under a national authority we should free ourselves from the habit of thinking of a park as necessarily a block of land, and draw up our planning scheme so that the reserved cliff lands shall be a strip as long as possible and no wider than is necessary. In this way the area will be much less than that of some of the proposed inland parks.

It is not, however, desirable that all the reserved cliff lands should all be under the central or national authority, for certain parts are of special importance to neighbouring seaside resorts. Thus the downs of Beachy Head are of great social and economic importance to Eastbourne, as the Corporation have already recognised, and the same considerations apply to the cliffs on the east side of Torquay. In many cases the proper destiny of the cliff lands adjoining seaside resorts appears to be their acquisition by the Local Authorities under the powers which have now been conferred upon them. What part of the cliff lands then is specially suitable for national as distinguished from regional reservation? It is remarkable that each principal consideration points in the same direction, namely to the cliffs of Cornwall. Thus if we consider the fine and lofty cliffs of the North Yorkshire coast we have to take account of the fact that only in summer is the climate attractive to visitors. Purbeck Island on the Dorset coast, with a milder climate, has wild and beautiful cliff lands, but being adjacent to Bournemouth and Poole is the special concern of these populous twin towns.

Cornwall has no great towns nor any very populous watering place. This county, moreover, has the warmest winter climate in England and the longest winter day, and in spring time the cliff lands are carpeted with wild flowers whose massed brilliance of colour rivals the flowery pastures of the Alps. The extensive cliffs of eruptive, metamorphic and ancient sedimentary rock are preeminent in all England for the quality of their substance. The massive granite cliffs of the Land's End look like ramparts piled up against the sea by the hands of giants, the serpentine rock of the Lizard promontory is of rugged solidity and varied colour, the hard slate rock of the cliffs from Boscastle to Tintagel has a deep blue tint which carries the chord of colour of sky and sea down to a depth of tone unequalled on the coasts of the other counties.

These three stretches of Cornish cliffs are specially suited for the beginnings of a national reservation of cliff lands to be acquired with the aid of funds from the Exchequer and placed under the national park authority.

No decision, however, has yet been taken as to the establishment of a national park authority, and the future action of Local Authorities with reference to cliff lands is still uncertain. Therefore the hope of greater things to come should not be allowed to interrupt the efforts of private persons to secure view-points on the cliffs of all parts of the English coast for the National Trust, in whose hands the amenities will be safeguarded for all time.

VIII. THE NEW FOREST AND THE FOREST OF DEAN AS NATIONAL PARKS

Cliff lands, woodlands and mountains are the three types of English scenery which it is of special importance to preserve on a scale sufficient to fulfil the requirements of a national park, that is to say with an area large enough to provide a complete environment of spontaneous Nature where no signs of Man's contrivance obtrudes upon the eye. A square of 10 miles side, comprising 64,000 acres, is sufficient in open country, and a smaller area in woodland.

There are two remaining Royal Forests large enough for the purpose, the New Forest and the Forest of Dean. The former has an area of 65,000 acres, the latter 28,000 acres when we add the adjacent Crown Woods. These two forests, together with smaller areas of Crown Lands were long since exchanged by the sovereign for the annuities of the Civil List and thus became national property. By a strange irony of circumstance, however, these national properties are now much less subject to popular control than are properties in private ownership, for whereas the Town and Country Planning Act gives Local Authorities considerable control over the use and development of private property, they have little or no voice in respect of national property, which is strictly guarded as a bureaucratic monopoly by one or other of several watertight Government departments.

The time has certainly come for the constitution of a Board of Scenery, on which the learned societies should be represented, charged with the administration of national parks and supervision of the planning schemes not only of Local Authorities but also of government departments. The latter are in special need of co-ordination in matters relating to scenic amenity.

6-2

The Crown Woods of the New Forest and the Forest of Dean, formerly administered by the Commissioners of Woods and Forests, were handed over to the Forestry Commissioners in 1924. The Commissioners are charged with the general duty of promoting the interests of forestry, the development of afforestation and the production and supply of timber in Great Britain. Including areas which have been purchased, the Commissioners hold 930,000 acres of which the New Forest and the Forest of Dean together constitute only one-tenth part. In relation to the main function of the Forestry Commission, namely the production of timber, these historic woodlands are of small importance. In a minute submitted to the National Park Committee, under the heading "Methods proposed to preserve the amenity of the [New] Forest"[1] the Commissioners state that the pine is an unsuitable tree. This is a fundamental fact in relation to the historic character of this typical woodland of the English plain to which only broad-leaved, hardwood trees properly belong. Out of consideration for amenity the Commissioners have decided to curtail the planting of conifers in areas suitable for hardwood trees, but nevertheless declare their intention of continuing to plant these unsuitable trees in other parts of the forest. If, however, the true historic character of the forest is to be preserved or restored, the woods of oak and beech should be offset and framed by a background of open heath, not camouflaged by alternating plantations of pine. The partial con-

[1] Appendix V of the Report of the Committee published April, 1931, pp. 113–16.

cession of profit-earning is doubtless evidence of a measure of sympathy on the part of the Commissioners, but still more is it evidence of the entire futility of a government policy which requires the New Forest and the Forest of Dean to be dealt with by compromise. However much the principle of give and take may be advocated in the conduct of public affairs in general, it is an entirely unpractical policy in the preservation of natural beauty, which depends upon harmonious grouping and therefore upon the complete avoidance of incongruous forms.

This single point, the planting of conifers in the New Forest, must suffice in a brief summary to illustrate the argument that amended instructions should be given to the Forestry Commissioners whereby they will no longer be required to take account of silvicultural value in the New Forest and the Forest of Dean, but be called upon to administer these beautiful and characteristic woodlands solely as national parks to be visited by the people of England both for the enjoyment of natural beauty and the cultivation of the historic sense.

One point remains to be mentioned with reference to the Forest of Dean. Here in addition to the beauty of woodland is the exquisite valley of the Wye, a river gorge in mountain limestone, one of the distinctive types of natural scenery which it is particularly important to protect from sophistication.

IX. THE MOORS OF THE PENNINE CHAIN

The moors of the Pennine Chain, wide uplands of unenclosed country, are particularly attractive to ramblers from the towns in the holiday season of summer when the landscape is carpeted with the rich colour of the heather blossom which imparts delicious fragrance to the bracing air. The number of ramblers, already very great, grows rapidly year by year and is largely recruited from the population of the neighbouring industrial centres. But on the grouse moors the interests of the owners are in sharp conflict with the requirements of the ramblers. The shooting rights are of great monetary value, and the question has consequently arisen whether some district of the Pennine Chain should be constituted a national park in the full sense, that is to say under a national authority and for the benefit of the people of all parts of England. The drawback is that where a broad and open type of landscape stretches illimitably before us the urge to vary the field of excursion and explore fresh ground is almost irresistible, so that a restricted area with artificial boundaries might not be permanently popular as a national park.

Here and there, however, in the Pennine region the council of some great industrial town may decide to secure free access to a neighbouring moor, thus constituting a regional as distinguished from a national park.

X. A NATIONAL PARK IN THE LAKE DISTRICT

England has only one region which displays the grandeur of mountain forms, the Lake District. For a full understanding of its advantages as the site of a national park we must examine its position on the map of Great Britain. Remarkably central as between south and north, it stands towards the west, the flank of the island where scenery is of the grander type. The long lakes which radiate in all directions from the central dome are on the same ample scale as the lochs of Scotland. The mountain masses of Skiddaw, Scafell and Helvellyn present bolder peaks than are seen in the mountain plateau of the Scottish Highlands, and rival in picturesqueness the Welsh ranges of Snowdonia. Moreover, the shores of the lakes and the lower dales have the luxuriant foliage of the English landscape and the enrichment of our old rustic architecture. The district which thus so happily combines the scenic charm of the three national countries of Great Britain happens to be on the scale best suited to the purpose of a national park, large enough but not too large for the excursions of a pedestrian and with a well-marked natural frontier, the outer rim of the "Cumbrian Dome", from which the rambler enjoys the wide prospect of the plain and then returns contentedly to his place of sojourn among the lakes and mountains.

In relation to residence the sub-regions of the Lake District are defined by contour lines of altitude, that part of the scenery which is unenclosed and untouched by the hand of man being the Fell country above 1000 feet. The lake shores and lower dales need careful provisions under the powers of the Town and Country Planning Act for the preservation of the picturesque amenities of a residential and an agricultural area. The several groups of the open fells should undoubtedly be placed without further delay under the aegis of a national park authority with powers to secure them not only against the common vulgarities of advertisements and litter but the more serious, because more lasting, encroachments of motor roads and pylons.

As a tourist resort the Lake District is not restricted to any single holiday season. The mild climate of the west brings soft air in spring when wild flowers bloom in profusion and the unfolding leaves display their various tints of green. Summer is pleasant for rambling on the hills, in autumn the gold and bronze of wood and fell are imaged in the magic mirror of the lakes, and in winter the summits of the mountains are silhouetted in silver against the sky.

The charms of the Lake District have exercised so strong an attraction that devotees of natural beauty have sought it from the earliest days when educated taste turned towards the wilder aspects of Nature. Here by Grasmere Wordsworth kindled the torch which first lit up the temple of Nature for our worship. Then Ruskin by the shore of Coniston kept the flame alight, and to these, more perhaps than to any other prophets of the century now past, we owe the faith, now beginning to be manifest, that the beauties of Nature are a source of inspiration unclouded by intellectual error and available for all men.

INDEX

Printed in the United States
By Bookmasters